Routledge Revivals

Entropy in Urban and Regional Modelling

First published in 1970, this groundbreaking investigation into *Entropy in Urban and Regional Modelling* provides an extensive and detailed insight into the entropy maximising method in the development of a whole class of urban and regional models.

The book has its origins in work being carried out by the author in 1966, when he realised that the well-known gravity model could be derived on the basis of an analogy with statistical, rather than Newtonian, mechanics. Subsequent investigation demonstrated that the entropy maximising method stems from an even higher level of generality, and the beginning of the book is devoted to an account of its importance and use as a general modelling tool.

This reissue will be welcomed by a range of students and professionals from fields as diverse as urban and regional studies, economics, geography, planning, civil engineering, mathematics and statistics.

Entropy in Urban and Regional Modelling

A. G. Wilson

Routledge
Taylor & Francis Group
ROUTLEDGE REVIVALS

First published in 1970
by Pion Ltd

This edition first published in 2011 by Routledge
2 Park Square, Milton Park, Abingdon, Oxon, OX14 4RN

Simultaneously published in the USA and Canada
by Routledge
711 Third Avenue, New York, NY 10017

Routledge is an imprint of the Taylor & Francis Group, an informa business

Publisher's Note
The publisher has gone to great lengths to ensure the quality of this reprint but points out that some imperfections in the original copies may be apparent.

Disclaimer
The publisher has made every effort to trace copyright holders and welcomes correspondence from those they have been unable to contact.

A Library of Congress record exists under ISBN: 0850860210

ISBN 13: 978-0-415-69564-0 (hbk)
ISBN 13: 978-0-203-14260-8 (ebk)
ISBN 13: 978-0-415-69631-9 (pbk)

Entropy in urban and regional modelling

A.G.Wilson

Pion Limited, 207 Brondesbury Park, London NW2 5JN

Library edition SBN 85086 021 0
Student edition SBN 85086 022 9

Set on IBM 72 Composers by Pion Limited, London.
Printed in Great Britain by J.W.Arrowsmith Limited, Bristol.

Preface

This book has its origins in some work I was doing in 1966 when I realised that the well-known gravity model could be derived on the basis of an analogy with statistical mechanics rather than Newtonian mechanics. Since then, I have realised that the entropy maximising method has roots at a higher level of generality and an account of the method as a general model building tool is presented in Chapter 1. I have also realised that it is useful to apply it to the development of a wide class of urban and regional models, and this application forms the main body of the book.

Since research workers in many disciplines are interested in urban and regional models, I hope that it will find readers in fields as different as economics, geography, planning, civil engineering, mathematics and statistics, and the newer, more specialised disciplines concerned with urban and regional studies such as regional science. It will probably be most useful to postgraduate students and research workers, but it should also have a role in advanced undergraduate teaching and for practitioners in various agencies of government. I myself have used material from the book, for example, in teaching part of a course in applied mathematics at an undergraduate level, and I see no reason why it should not interest undergraduates in other departments, especially social science departments which are trying to develop, in part at least, a more quantitative approach.

Some of the material published in this book is reworked from papers published in various journals. I would like to acknowledge the editors of the following journals in this respect: *Operational Research Quarterly*, for some material in chapter 1; *Transportation Research*, for some material in chapter 2; *Geographical Analysis*, for some material in chapter 3; *Papers*, *Regional Service Association*, for some material in chapters 4 and 7; and the *Journal of Regional Science*, for some material in chapters 4 and 5.

I have benefited considerably from discussion with colleagues in the Ministry of Transport and the Centre for Environmental Studies, and with students and colleagues in lectures and seminars in many University Departments throughout this country and abroad. It is invidious to pick out particular names but I would like to mention my particular debt to Tony Blackburn, of Harvard University, and Britton Harris of the University of Pennsylvania, who studied my earlier papers in this field with great care, and whose criticisms forced me to clarify many of my ideas. I also received much encouragement from Walter Isard and his colleagues in the Regional Science Department in Philadelphia, who generated much useful interaction by asking me to present papers on entropy maximising methods to Regional Science Association Conferences. Finally, I would like to thank Angela Hall and Sheila Jamieson who have patiently typed and retyped a difficult manuscript. I alone, of course, am responsible for any errors of all kinds which remain.

A.G.Wilson
Professor of Urban and Regional Geography, University of Leeds

To Christine

Contents

1

What is entropy?

1.1 Introduction

Urban and regional models are of interest for two main reasons. Firstly, model building is at the root of all scientific activity, and urban and regional modelling is part of an attempt to achieve a scientific understanding of cities and regions. Secondly, a variety of severe urban and regional problems exist, and associated planning activity has become increasingly important: urban and regional modelling is a part of the advance on this front also.

The urban and regional scientist faces a number of severe theoretical problems. His activity is often a multidisciplinary one in the sense that he needs to use concepts from several disciplines—economics, geography, sociology, and so on. The concept of entropy has, until recently, been used primarily in the nonsocial sciences. This book is an attempt to show that it has a useful and valuable role in one branch of the social sciences—the study of urban and regional systems—and, in passing, the account may stimulate people in other fields also. It is hoped that 'entropy' enables the social scientist to tackle some of his basic problems in a fruitful way, and thus to make progress which might not be possible so easily with more orthodox tools.

One of the initial problems is that entropy as a concept can be used in several ways, and so the first task, in Section 1.2 below, is to outline three views of the concept. In Section 1.3, we shall discuss in general terms how the concept can be used: as a theory building and as an interpretive aid, and to assist in the study of system dynamics. Various examples are then discussed: transport flows (urban and regional) of various kinds in Chapters 2 and 3; location models in Chapter 4; utility maximising systems in Chapter 6. Some technical problems concerned with modelling can be solved using entropy maximising procedures, and these are shown in Chapter 5. There is a general discussion of the relation of the concept to social physics and general systems theory in Chapter 7. A number of conclusions are drawn in Chapter 8.

1.2 Three views of entropy

1.2.1 The relationship of entropy to probability and uncertainty

A 'system of interest' in the context of this book is any urban or regional system which it is useful to study. Many examples will be discussed. It is necessary to begin by defining and specifying a *state* of the system of interest. What information do we need to be given to specify fully a system state? This question can be partly answered by giving one or two examples, though we shall see later that it is a far from trivial question. In physics, and in particular in classical physics, the state of a gaseous system is fully specified by the coordinates and velocities of each particle in the gas at any time. Such a system should be analysed from first principles using the techniques of Newtonian classical physics. However, such techniques (which can be described as *microanalytic* for a gas) prove too difficult to handle in many common situations when there are of the order of more than 10^{23} particles involved. In fact, they are pretty difficult when more than two particles are involved. A branch of physics, known as statistical mechanics, developed in order to study such systems, and new techniques emerged which enabled the physicist to explain and predict certain *macroproperties* of the system to a desired degree of accuracy without having to explain (at the *microlevel*) the behaviour of each individual particle. Such macroanalytic techniques are related to microsituations using the concept of entropy (cf. Jaynes, 1957).

Now, consider an equivalent example from the social sciences. Take a system with a fixed spatial distribution of numbers of workers in residences, and a fixed spatial distribution of number of jobs. The system of interest is taken to be the flow of workers from residences to jobs. How would we now specify a *state* of the system? Simply: it would be necessary to state, for each individual worker, his residential zone and his workplace zone. In a large city, the task of predicting the behaviour of each individual in this respect is too great (as with the corresponding situation in the study of gases). Once again, we may only be interested in certain macroproperties of the system, and it turns out that such properties can be explained and predicted by using statistical techniques which use the concept of entropy.

These examples, then, illustrate the kind of situation where we need to use the concept of entropy, and we now have to face up to the basic question: what is entropy? The second example above, the 'journey-to-work system', will be used to illustrate the concept by relating it to the difficult, but perhaps better understood, concepts of probability and uncertainty.

In our example, the system of interest can be specified by an origin-destination table, as shown in Figure 1.1.

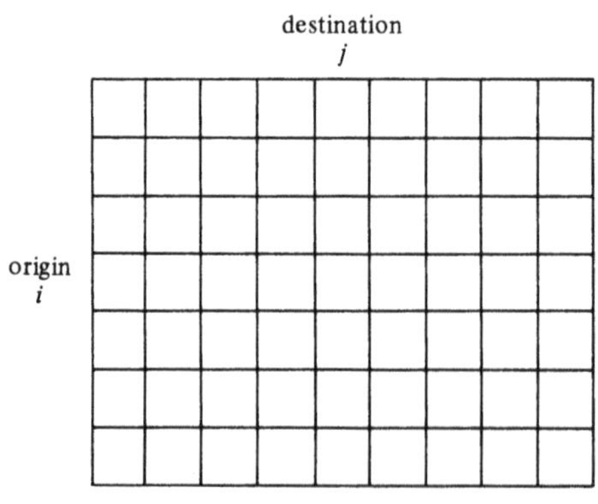

Figure 1.1. Origin-destination table.

A state of the system is an *assignment* of individual workers to the origin–destination table *which is in accord with any constraints*, such as the requirement, mentioned above, that the distribution of numbers of worker residences and jobs is assumed to be given. This is a complete description of the system of interest. We must also define a *distribution* which is the macroproperty of the system we wish to estimate by statistical means. In this case, a distribution is a set of numbers, one for each origin–destination pair, and one such number is the total number of people who travel from origin i to destination j. Thus each element, or 'box', of the origin–destination matrix in Figure 1.1 can be considered to contain two components: firstly, a *set* of individual workers who travel from i to j (the total collection of such sets being called an assignment of individuals for the whole system, and an assignment defines a single *state* of the system); and, secondly, a total T_{ij} of individuals travelling from i to j (the set of such numbers being called a *distribution* of the system).

It can easily be seen that there are many states which give rise to any particular distribution. Suppose we make the assumption that any state of the system occurs with equal probability. Then, we can find the most probable distribution by calculating the set of T_{ij}'s which has the greatest number of states associated with it. This calculation can be carried out without any knowledge being needed of the particular individuals. We shall now do this as a preliminary to a further discussion of entropy, probability, and uncertainty.

Before proceeding with the calculation, it may be possible to clarify this crucial part of the discussion for the reader by considering the concept of a system state in more depth. A state of the system has been defined so far as a complete description of the system of interest. Let us call such a description a *microstate* description. Then a *distribution* can be seen as a *macrostate* description, as it is less than complete in terms of information, and, as we have seen, many microstates can give rise to the same macrostate. We have also discussed constraints: for example, that the locational distribution (as distinct from journey-to-work *trip* distribution) of worker residences and jobs is exogenously given, so that any trip distribution is constrained by the given locational distribution. We shall see shortly that it is also convenient to impose a total-transport-cost constraint. Thus the specification of the locational-distribution total cost is a higher-level macrostate description. It obviously contains less information than the trip-distribution macrostate description, and so many trip distributions can give rise to the same locational distribution. The interrelationships of these state definitions are shown symbolically in Figure 1.2. Each 'box' represents a state. This figure illustrates the argument of the preceding paragraph: if all microstates are equally probable, we can find the most probable trip distribution by calculating the number of microstates associated with each distribution, subject to any relevant constraints; and in Figure 1.2 terms the constraint qualification means that we are only interested in trip-distribution macrostates which produce some exogenously given locational-distribution macrostate. This structure is common in the development of entropy-maximising models, though the details can obviously be changed so as to be in accord with the model builders assumptions and hypotheses. (In an equivalent situation in physics, of the gaseous system referred to earlier, microstates correspond to the coordinate-velocity description; the middle-level macrostate of Figure 1.2 corresponds to the numbers of particles in each energy level; and the upper-level macrostate corresponds to an exogenously given total system energy and total number of particles.)

We can now proceed with the calculation.

It has been shown (Wilson, 1967a) that three constraints are needed to generate a good estimate of T_{ij}; first we define the following variables:

T_{ij} is the number of individuals living in i and working in j (to be estimated);
O_i is the total number of workers who live in i (given);
D_j is the total number of jobs in j (given);
c_{ij} is the cost of travelling from i to j (given);
C is the total expenditure on travel to work (given).

Then the *constraints* on T_{ij} (which restrict the number of assignments giving rise to a distribution) can be written:

$$\sum_j T_{ij} = O_i \tag{1.1}$$

$$\sum_j T_{ij} = D_j \tag{1.2}$$

$$\sum_i \sum_j T_{ij} c_{ij} = C \ . \tag{1.3}$$

Now, we want to find the matrix $\{T_{ij}\}$ which has the greatest number of states, say $W(\{T_{ij}\})$, associated with it, subject only to the constraints (1.1)–(1.3). We can obtain the number of states which give rise to a matrix $\{T_{ij}\}$ as follows. Suppose T is the total number of workers (that

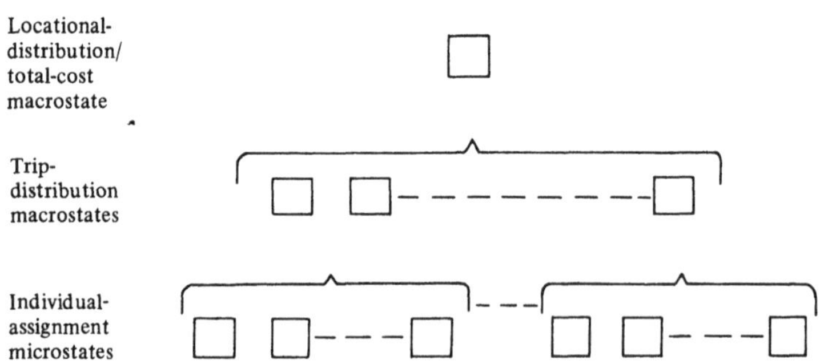

Figure 1.2. Macrostates and microstates.

is, $T = \sum_i \sum_j T_{ij}$). How many assignments of individual workers to 'boxes' of Figure 1.1 give rise to $\{T_{ij}\}$? Firstly, we can select T_{11} from T, T_{12} from $T-T_{11}$, etc., and so the number of possible assignments, or states, is the number of ways of selecting T_{11} from T, ${}^{T}C_{T_{11}}$, multiplied by the number of ways of selecting T_{12} from $T-T_{11}$, ${}^{T-T_{11}}C_{T_{12}}$, etc. Thus

$$W(\{T_{ij}\}) = {}^{T}C_{T_{11}}\,{}^{T-T_{11}}C_{T_{12}}\,{}^{T-T_{11}-T_{12}}C_{T_{13}} \cdots . \tag{1.4}$$

So, explicitly,

$$W(\{T_{ij}\}) = \frac{T!}{T_{11}!(T-T_{11})!} \cdot \frac{(T-T_{11})!}{T_{12}!(T-T_{11}-T_{12})!} \cdot \cdots = \frac{T!}{\prod_{ij} T_{ij}!} , \tag{1.5}$$

and a little thought shows that this result is independent of the order in which the 'boxes' of Figure 1.1 are considered.

We now maximise $W(\{T_{ij}\})$ subject to Equations (1.1) to (1.3) in order to find the most probable $\{T_{ij}\}$. In fact, of course, any monotonic function of $W(\{T_{ij}\})$ can be used to give the same result, and for convenience we maximise $\ln W(\{T_{ij}\})$ subject to Equations (1.1)–(1.3). This gives T_{ij} as

$$T_{ij} = A_i B_j O_i D_j \exp(-\beta c_{ij}) , \tag{1.6}$$

where

$$A_i = \frac{\exp(-\lambda_i^{(1)})}{O_i} = \left[\sum_j B_j D_j \exp(-\beta c_{ij})\right]^{-1} \tag{1.7}$$

and

$$B_j = \frac{\exp(-\lambda_j^{(2)})}{D_j} = \left[\sum_i A_i O_i \exp(-\beta c_{ij})\right]^{-1} , \tag{1.8}$$

where β is the Lagrangian multiplier associated with (1.3) and $\lambda_i^{(1)}$ and $\lambda_j^{(2)}$ are the sets of Lagrangian multipliers associated with Equation (1.1) and (1.2), respectively. (It is more convenient to use A_i and B_j, transformations of the multipliers, as shown.) The proof is given in an earlier paper (Wilson, 1967a) and in Chapter 2 below. The derivation relies on the use of Stirling's approximation:

$$\ln T_{ij}! = T_{ij} \ln T_{ij} - T_{ij} . \tag{1.9}$$

The estimate of $\{T_{ij}\}$ given by Equations (1.6)–(1.8) has been used as an empirical or phenomenological estimate, without this derivation, for some years, and is in reasonable accord with reality. Many refinements are possible; they are discussed elsewhere (Wilson, 1967a, 1969a), and a more detailed discussion for present purposes is postponed to Chapter 2.

What can we now say about entropy, probability, and uncertainty? This derivation parallels a technique in statistical mechanics known as the use of the microcanonical ensemble (see Wilson, 1969b, and below). In statistical mechanics, the equivalent of our $\ln W$ above is *defined* to be the *entropy of the system.* We have seen that it is the log of the *probability* that a distribution occurs, so entropy is monotonically related to probability defined in this way.

It can be seen just as easily that entropy can be related to *uncertainty.* Our uncertainty relates to states of the system (and, in particular, to microstates, in Figure 1.2 terms). We have said that we are interested in distributions only and that the most probable distribution is that with the greatest number of microstates giving rise to it. Thus the distribution corresponds to the position where we are most uncertain about the microstate of the system, as there are the largest possible number of such states and we have no grounds for choosing between them. Thus another way of stating our maximum-probability principle would be to say that we have no grounds for choosing a distribution which reduces our admitted uncertainty about microstates. If we had such grounds, they would imply a rewriting of the constraint equations, and this would transform the problem into another one which would satisfy the rule just stated.

It is of some interest that the ideas discussed in this section were being developed in a number of ways independently by a variety of authors in different parts of the world. Murchland (1966) showed, using a method of Samuelson (1947), that the gravity model could be derived from a maximisation problem, and he used the mathematical programming formulation thus achieved to prove some uniqueness theorems. The idea of maximising a function of the form of $\ln W$ can be seen in the work of Schneider (1967), though he used the method for a slightly different purpose. The maximisation of $\ln W$ to derive the gravity model, based on analogies with statistical mechanics, have been pursued in various ways by Spurkland (1966), in the context of the Oslo Transportation Study, by Tomlin (1967) (also discussed by Tomlin and Tomlin, 1968), by Wilson (1967a), as described above, by Loubal (1968), and by Sasaki (1968). Neidercorn and Bechdolt (1969) have more recently attempted to derive a form of gravity model from what they call a utility-maximising procedure. This involves identifying the concept of entropy with that of utility in a way which is rarely possible. The use of entropy-maximising methodology in the analysis of utility-maximising systems is discussed in Chapter 6 below. It is also instructive to investigate this 'probabilistic' mathematics in detail from the viewpoint of the theory of ensembles (Gibbs, 1902). This is done in Appendix 2.

1.2.2 The entropy of a probability distribution

This section presents the ideas of Jaynes (1957) and will enable us to offer an alternative interpretation of 'entropy'. Let x be a random variable which can take values $x_1, x_2, ..., x_n$ with probabilities $p_1, p_2, ..., p_n$. The probabilities are not known. All we know is the expectation value of some function $f(x)$:

$$\sum_i p_i f(x_i) = E[f(x)] \ . \tag{1.10}$$

We also know that

$$\sum_i p_i = 1 \ . \tag{1.11}$$

Given this information only, what is our best estimate of the probability distribution p_i?

Jaynes writes: "Just as in applied statistics the crux of a problem is often the devising of some method of sampling that avoids bias, our problem is that of finding a probability assignment which avoids bias while agreeing with whatever information is given. The great advance provided by information theory lies in the discovery that there is a unique, unambiguous criterion for the 'amount of uncertainty' represented by a discrete probability distribution, which agrees with our intuitive notions that a broad distribution represents more uncertainty than does a sharply peaked one, and satisfies all other conditions which make it reasonable". This measure of the uncertainty was given by Shannon (Shannon and Weaver, 1949) as

$$S(p_1, p_2, ..., p_n) = -k\sum_i p_i \ln p_i \ , \tag{1.12}$$

and this is defined to be the entropy of the probability distribution $p_1, p_2, ..., p_n$. The proof that this is a unique, unambiguous measure of uncertainty is given in Appendix 1 below.

Jaynes then writes: "It is now evident how to solve our problem; in making inferences on the basis of partial information we must use that probability distribution which has maximum entropy subject to whatever is known. This is the only unbiased assumption we can make; to use any other would amount to arbitrary assumption of information which by hypothesis we do not have".

Thus, to solve the problem posed above, we simply have to maximise entropy in Equation (1.12), subject to Equations (1.10) and (1.11), which represent what we know. This gives

$$p_i = \exp[-\lambda - \mu f(x_i)] \tag{1.13}$$

where, from Equation (1.11),

$$e^{\lambda} = \sum_i \exp[-\mu f(x_i)] \tag{1.14}$$

and λ and μ are the Lagrangian multipliers associated with Equations (1.11) and (1.10), respectively. This result can easily be generalised to the case where the expectation values of several functions of x_i are known (as given by Jaynes, 1957, and described by Wilson, 1969b).

We now have to show that this measure of 'uncertainty' is the same as that introduced in Section 1.2.1 above. Define

$$p_{ij} = \frac{T_{ij}}{T} \ . \tag{1.15}$$

Then, from Equation (1.5),

$$\ln W = \ln T! - \sum_{ij} (T_{ij} \ln T_{ij} - T_{ij}) \tag{1.16}$$

after the use of Stirling's approximation for $T_{ij}!$ This can be written in terms of p_{ij} as

$$\begin{aligned} \ln W &= \ln T! - \sum_{ij} Tp_{ij} [(\ln p_{ij} + \ln T) - Tp_{ij}] \\ &= \ln T! - T\sum_{ij} p_{ij} \ln p_{ij} - (T \ln T - T)\sum p_{ij} \\ &= (\ln T! - T \ln T + T) - T\sum_{ij} p_{ij} \ln p_{ij} \ . \end{aligned} \tag{1.17}$$

Hence, maximising

$$S(p_{11}, p_{12}, ..., p_{nn}) = -\sum_i \sum_j p_{ij} \ln p_{ij} \tag{1.18}$$

subject to Equations (1.1)–(1.3) [which can be expressed as constraints on the p_{ij} using Equation (1.5)] will lead to an estimate of T_{ij} which is the same as that given in Section 1.2.1 above. $\ln W$ and S are, in fact, linearly related.

However, the differences between the two definitions are not entirely trivial, and the second is to be preferred for several reasons. Firstly, any derivation based on the $\ln W$ type of definition relies on Stirling's approximation in the derivation of the answer, while the second type of definition does not. (However, this is not a very important reason: we only use the derivation of Stirling's approximation, and the derivatives coincide in the two cases[(1)]). Secondly, and more important, the second definition of entropy is more flexible and useful than the first. The difference between the definitions is analogous to the objective and subjective views of probability. Quoting, and paraphrasing, Jaynes (1957): "The objective view is that probability is always capable of measurement by observation of frequency ratios in a random experiment; the subjective view regards probabilities as expressions of human ignorance; the probability of an event is merely a formal expression of our expectation that the event will, or did, occur, based on whatever information is available. To the subjectivist, the purpose of probability theory is to help us in forming plausible conclusions in cases where there is not enough information available to lead to certain conclusions; thus detailed

(1) I am grateful to Britton Harris for pointing this out to me.

verification is not expected. The test of a good subjective probability distribution is: does it correctly represent our state of knowledge as to the value of x?". It can easily be seen that the first definition (in the previous section) is essentially objective and the second (in this section) is essentially subjective.

It is not necessary to try to reduce everything to either the objectivist or subjectivist school of thought; both views are useful at appropriate times. It is important not to mix the two views, however. There is also a converse to this analysis: entropy is not a Platonic concept associated with systems. If there is any desire to use the concept of entropy, then (a) it should be made quite clear how it *could* be measured, or (b) the probability distribution associated with some property of the system should be revealed as that giving rise to this system entropy.

A more detailed discussion of the types of probability distribution which can be defined for systems, related to the corresponding definition of entropy, is given in Appendix 2. Further, in Appendix 4, an alternative method of estimating T_{ij} or p_{ij}, the Darwin–Fowler method, is discussed (cf. Wilson, 1967c).

1.2.3 Entropy and Bayesian statistics

The Bayesian statistician is essentially a subjectivist, as he is concerned with degrees of belief and estimating probability distributions which are in the best possible accord with the evidence. However, the objectivist–subjectivist line is a fine one and the book by Lindley (1965) (on which the development of this section is based) gives a frequency interpretation of degrees of belief. Bearing all this in mind, let us study Bayesian methods of statistical inference and we shall see that it can be related to entropy, and will thus give us a third view of this concept.

Let x be a *random variable* which can take values $x_1, x_2, x_3, \dots$. A *random sample* of size n is defined to be a set of independent random variables, each of which has the same distribution as x. Let $p(x_i|\theta)$ ($= p_i$, for short) specify the distribution, where θ is a parameter of the distribution. Let H denote our state of knowledge before the sample is taken. Then θ will have a distribution dependent on H which can be written $\pi(\theta|H)$. Let the vector $\hat{\mathbf{x}} = (\hat{x}_1, \hat{x}_2, \dots)$ be a random sample. Then its density can be written

$$\mathrm{f}(\hat{\mathbf{x}}|\theta, H) = \prod_{i=1}^{n} p(\hat{x}_i|\theta) \tag{1.19}$$

where the right-hand side can be written as a product because the x_i's have been assumed to be independent. Then the new distribution of θ given to us by the random sample $\hat{\mathbf{x}}$ is given by Bayes's theorem as

$$\pi(\theta|\hat{\mathbf{x}}, H) = \frac{\mathrm{f}(\mathbf{x}|\theta, H)\pi(\theta|H)}{\pi(\hat{\mathbf{x}}|H)} . \tag{1.20}$$

We can obtain the maximum-likelihood estimate of θ, say $\hat{\theta}$, by finding the θ which maximises $f(\hat{\mathbf{x}}|\theta, H)$, which can be written $f(\hat{\mathbf{x}}|\theta)$ for short, with H understood. It is often more convenient to work with the log likelihood, called $L(\hat{\mathbf{x}}|\theta)$ and obtained by taking logs in Equation (1.19):

$$L(\hat{\mathbf{x}}|\theta) = \sum_{i=1}^{n} \ln p(\hat{x}_i|\theta) \quad . \tag{1.21}$$

As the sample size n increases, the value of θ, $\hat{\theta}$ obtained by maximising $L(\hat{x}|\theta)$ will, under appropriate conditions, tend to the true value, and, further, we have from the law of large numbers (Theorem 3.6.3, as discussed in Section 7.1, of Lindley, 1965)

$$\lim_{n\to\infty}[n^{-1}L(\hat{\mathbf{x}}|\theta)] = E[\ln p(x|\theta)] \quad . \tag{1.22}$$

However,

$$\begin{aligned} E[\ln p(x|\theta)] &= \sum p(x_i|\theta)\ln p(x_i|\theta) \\ &= \sum p_i \ln p_i \quad . \end{aligned} \tag{1.23}$$

This is the negative of what we have defined as the entropy of a probability distribution in Section 1.1.2 above. This means that, if we choose the form of the function p by maximising its entropy, we are assuming the form which minimises the likelihood function. This is another way of stating our previous result that the probability distribution which maximises entropy makes the weakest assumption which is consistent with what is known. Of course, the entropy-maximising procedure would normally also give parameter estimates (as the parameters would be Lagrangian multipliers which could be estimated by solving the constraint equation). Thus we have a puzzle to resolve: how Bayesian maximum-likelihood procedures and entropy-maximising procedures can be consistent, and how, in certain circumstances, they can complement each other. This puzzle will be resolved in Section 1.3.5 below.

An alternative approach relating entropy-maximising procedures using contingency tables has been given by Hyman (1969), following Tribus, Evans, and Crellin (1964).

1.3 The application of the concept of entropy[2]

1.3.1 Summary of types of application

The rest of this book will be concerned with detailed examples of the application of the concept of entropy in urban and regional modelling, that is, in *hypothesis development*, or theory building. ('Model' and 'hypothesis' are used synonymously, and a theory is a well-tested hypothesis.) We have already seen in the example sketched in

(2) I am indebted to A. J. Blackburn for stimulating discussions on topics discussed in this section, though of course he bears no responsibility for the views expressed.

Section 1.2.1, and in the more general discussion of Section 1.2.2, that the entropy-maximising procedure can be used to develop hypotheses and thus will be the main type of application. The notion can be reversed, however: in cases where hypotheses can be developed by more conventional means, it will often be useful to exhibit the constraint equation which would give rise to the same hypotheses in an entropy-maximising procedure, as this will often facilitate *interpretation*. This is the second main application. Thirdly, we shall see that entropy-maximising procedures facilitate the study of system dynamics. These three types of application are discussed in Sections 1.3.2, 1.3.3, and 1.3.4, respectively, below. In Section 1.3.5 we shall use our knowledge thus gained to review the connection between entropy-maximising procedures and statistical analysis. Finally, in Section 1.3.6 we shall mention a number of other possible applications which are not considered in detail in this book.

1.3.2 Hypothesis generation

The general rule for generating hypotheses, which is exhibited by the examples of Sections 1.2.1 and 1.2.2, can be stated as follows. Set up the variables which define the system of interest, and write down the known constraint equations on these variables. Define the entropy of the system, either directly as in the transport example of Section 1.2.1, or by using an associated probability distribution as in Section 1.2.2 (where the two were shown to be equivalent). The variables can then be estimated by maximising the entropy subject to the constraints. Many, if not most, hypotheses thus generated could be produced by more conventional means. However, it will be seen in the examples presented in the rest of the book that, at the very least, the entropy-maximising procedure enables us to handle extremely complex situations in a consistent way, and past experience has shown that this sort of consistency is very difficult to achieve otherwise.

However, it should be emphasised (with Clough, 1964) that the hypotheses which are generated should be tested in the same way as hypotheses generated by any other procedure.

1.3.3 Interpretation of theories

When we construct models, hypotheses, or theories, it is often necessary to include terms which are difficult to interpret in any direct way. These are high-level theoretical concepts which are often well removed from possibilities of direct measurement. They may be the parameters of a model, such as β in the transport model given by Equation (1.6). Suppose (as was in fact the case) the model given by Equations (1.6)–(1.8) could be developed and used fruitfully without entropy-maximising procedures. It is then possible to write down the set of constraint equations which give rise to the same model, in this example Equations (1.1)–(1.3). The parameter β is then seen to be the Lagrangian multiplier associated with the constraint Equation (1.3) and the interpretation of this equation adds to our knowledge of the role β plays in the model.

1.3.4 The role of 'entropy' in the study of system dynamics

The concept of entropy is perhaps best known (if not, understood) by its presence in one of the many forms of the second law of thermodynamics: that entropy always increases. (The situation is even more confused if the equations entropy = uncertainty = chaos are added to this statement.) The second law arises from an equation of the form

$$ds = \mu dE + \mu \sum_k X_k \, dx_k \tag{1.24}$$

where ds is the change in entropy, dE the change in system energy, and X_k the external force corresponding to external coordinate x_k, and so

$$-\sum_k X_k \, dx_k$$

is the work done externally *on the system.* Thus the second law, expressed as

$$ds \geqslant 0 \; , \tag{1.25}$$

simply asserts that the system cannot receive more in energy than the amount of external work supplied. Thus, in this case, entropy could (as a useful measure of a quantity *objectively* defined in the system) be defined by Equation (1.24). This objective use of the concept, in contrast to the three essentially subjective uses of Section 1.2, can be considered a fourth 'view' of entropy.

There are examples where the concept can play an equally useful role in social systems. It is important, however, that the concepts corresponding to energy and work (such as utility and investment) are defined explicitly for the system of interest. Such examples, together with the beginning of an indication of why the entropy-maximising procedure gives an advantage over conventional statistical analysis in setting up dynamic models, are discussed in Appendix 4 of an earlier paper by the author (Wilson, 1969b), and in Chapter 7 below.

1.3.5 Review of interrelationship of entropy-maximising procedures and statistical analysis

It is now necessary and fruitful to return to the puzzle raised in Section 1.2.3: in effect, how can we reconcile the approach of the statistician and the entropy maximiser to the same problem? Let us explore this with a simple problem. Suppose x is a random variable with possible values x_i, and the probability of x having a value x_i is p_i. This can be written $p(x_i \mid \theta)$, where θ is a single parameter, as in the example of Section 1.2.3. We can assume that the entropy maximiser in this example expresses his knowledge of constraints on p_i in a single equation, as in

Section 1.2.2. This can be taken as Equation (1.10), which is repeated here for convenience:

$$\sum_i p_i \mathrm{f}(x_i) = E[\mathrm{f}(x)] \tag{1.10}$$

and the maximum entropy estimate of p_i is given by Equations (1.13) and (1.14) and can be written as

$$p_i = \frac{\exp[-\mu \mathrm{f}(x_i)]}{\sum_i \exp[-\mu \mathrm{f}(x_i)]} \tag{1.26}$$

and so p_i is a function of a single 'parameter' μ.

The statistician proceeds by assuming a number of appropriate functional forms for p, he uses data to obtain a maximum-likelihood estimate of the parameter $\theta, \hat{\theta}$, and he can carry out goodness-of-fit tests. He chooses the function which gives the best fit, and the corresponding θ.

The entropy maximiser obtains his estimate of p_i and can also carry out goodness-of-fit tests on his model predictions. If his predictions are not in accord with reality, he can change the function form of $\mathrm{f}(x_i)$ in Equation (1.10), provided he knows the expectation value of the new function, until he obtains better predictions.

If the entropy maximiser, as he tries new forms of function f, *does not know* the expectation value on the right-hand side of Equation (1.10), then he is in exactly the same position as the statistician. He must use a statistical procedure (such as a maximum-likelihood procedure) to estimate his parameter μ. In these circumstances, the statistician and the entropy maximiser should reach the same best estimate of both functional form and parameter.

Tribus (1968) shows how different distributions can be generated from different forms of constraint equations. The 'parameters' of distributions which are formed in this way are in this case Lagrangian multipliers (or functions of them) associated with the constraint equations, and so can be determined exactly by solving the constraint equations.

The entropy maximiser has the three potential advantages, however, over the statistician implied by the discussions of Sections 1.3.2–1.3.4 above: firstly, dealing with constraints rather than the distribution function directly may enable him to achieve consistency more easily in complex situations; secondly, he may be able to interpret the constraint equations and thus facilitate his understanding of the system; thirdly, he may be able to proceed more directly to a dynamic model using general principles of systems analysis.

Two generalisations should be noted. Firstly, the statistician may want to introduce additional parameters, and this is equivalent to the entropy maximiser introducing additional constraints. Secondly, a possible adjustment which can be made by either statistician or entropy maximiser is to the specification of possible values of the random variable x.

Clearly, if the x_i's are wrongly specified, predictions will be poor.

One final response should be made to a possible difficulty. It has been argued that when an entropy maximiser uses constraint equations to generate hypotheses or distribution functions *and* (in at least some cases) he does not know the numerical value of the expectation values in the equations, then he is unreasonably picking out particular expectation-value equations from a possible infinite number. Why should he pick out particular ones and not others? This question can be answered by the assertion that the functions of the random variables whose expectation values appear in the equations selected plays a special role in the system of interest. This is most dramatically illustrated by a common example from physics already referred to in Section 1.2.1: the microcanonical analysis of particles in gas. The constraint equation in this case concerns the energy of the system, and energy plays a special role because it is conserved. Thus, if an entropy-maximising model gives good predictions by reproducing complex real-world situations, the constraint equations should be susceptible to useful interpretations of this kind.

1.3.6 Other applications of the concept of 'entropy'

We have concentrated on the use of 'entropy' in theory-building and interpretation procedures, though we have mentioned one particular additional role—its use in expressing laws about system dynamics. These theory-building applications in urban and regional modelling are discussed in much more depth in other papers (Wilson, 1967a, 1969a, 1969b; Mogridge, 1969) and will be actively pursued in the rest of the book.

It is perhaps worth mentioning a number of other quite specific applications of the concept. Firstly, in information theory itself, the 'expected information' of a message is what we have defined to be entropy, and this measure can be used objectively to match information streams with channel capacities, and to solve certain encoding problems, in electrical engineering (see, for example, Goldman, 1955, Raisbeck, 1963, Tribus, 1968).

Secondly, Theil (1967) has used the concept of entropy, or expected information, to study loss of information when certain kinds of category aggregation arise in economic problems. It could similarly be used to study zone-aggregation problems. Further, economic applications are given by Murphy (1965). He pays particular attention to the subjective nature of expectations and associated probability distributions and shows how particular care has to be taken in social science applications of entropy.

Transport models: the theory of trip distribution, mode split, and route split

2.1 Introduction

In transport planning, a model is used to estimate trips between pairs of points (as an aid to network design and economic evaluation), and to simulate flows of links of the transport system (as an integral part of the model, and as an aid to traffic engineering). The model usually consists of four submodels concerned respectively with trip generation, trip distribútion, model split, and assignment (see, for example, Martin, Memmott, and Bone, 1961, Overgaard, 1966). This chapter is not concerned with trip generation: it assumes that estimates of trip ends are known (using the model, for example, of Wootton and Pick, 1967). Some of the results have implications for assignment, but we are mainly concerned with trip distributon and modal split. The application of entropy-maximising methods to these topics has been set out in two papers (Wilson, 1967a, 1969a), and this chapter will review the ground covered by these papers, with some additional comments and results.

It will be seen that the main advantage of entropy-maximising methods in the transport field, in the language of Chapter 1 above, is their ability to help in constructing models to represent complex phenomena, and, to a lesser extent, their use in the interpretation of the equations. We shall begin by giving a full discussion of the gravity model of trip distribution in Section 2.2 (which is a full exposition of the example introduced in Chapter 1). In Section 2.3 we shall show how to construct more general distribution models, and in Section 2.4 draw the modal-split implications. A discussion of the underlying hypotheses (Section 2.5) leads to an alternative modal-split model (Section 2.6) and to an examination of the concepts of composite impedance (Section 2.7) and route split (Section 2.8). The chapter concludes with a discussion of the results achieved (Section 2.9).

2.2 The gravity model of trip distribution

2.2.1 Derivation of the gravity model

The task of any trip distribution model is to estimate the number of trips between each pair of 'points'. (A 'point' in this case will be the centroid of a zone: it will be assumed that the study area of interest has been divided into zones in a suitable way.) We can, in the first instance, use the notation already introduced in Chapter 1. Choose a single purpose as an illustration, say the journey to work to fix ideas. Let T_{ij} be the number of (work) trips and c_{ij} the travel cost between zones i and j; let O_i be the total number of work-trip origins in i; and let D_j be the total number of work-trip destinations in j. A distribution model then estimates T_{ij} as a function of the O_i's, the D_j's, and the c_{ij}'s.

These variables could, of course, themselves be functions of other independent variables.

The simplest such model is the so-called *gravity model* developed by analogy with Newton's law of the gravitational force F_{ij} between two masses m_i and m_j separated by a distance d_{ij}:

$$F_{ij} = \gamma \frac{m_i m_j}{d_{ij}^2} , \tag{2.1}$$

where γ is a constant. The analogous transport gravity model is then

$$T_{ij} = k \frac{O_i D_j}{c_{ij}^2} , \tag{2.2}$$

using the variables defined above, where k is a constant and where travel cost is interpreted as 'distance'. This model has some sensible properties: T_{ij} is proportional to each of O_i and D_j and inversely proportional to the square of the distance between them. But the equation has at least one obvious deficiency: if a particular O_i and a particular D_j are each doubled, then the number of trips between these zones would quadruple according to Equation (2.1), when it would be expected that they would double also. To put this criticism of Equation (2.1) more precisely, the following constraint equations on T_{ij} should always be satisfied, and they are not satisfied by Equation (2.1):

$$\sum_j T_{ij} = O_i \tag{2.3}$$

$$\sum_i T_{ij} = D_j \ . \tag{2.4}$$

That is, the row and column sums of the trip matrix should be the numbers of trips generated in each zone and the number of trips attracted, respectively. These constraint equations can be satisfied if sets of constants A_i and B_j associated with production zones and attraction zones respectively are introduced. They are sometimes called balancing factors. Also, there is no reason to think that distance plays its part in the transport Equation (2.2) as it does in the world of Newtonian physics, and so a general function of distance is introduced. The modified gravity model is then

$$T_{ij} = A_i B_j O_i D_j \mathrm{f}(c_{ij}) \tag{2.5}$$

where

$$A_i = \left[\sum_j B_j D_j \mathrm{f}(c_{ij}) \right]^{-1} \tag{2.6}$$

and

$$B_j = \left[\sum_i A_i O_i \mathrm{f}(c_{ij}) \right]^{-1} . \tag{2.7}$$

The equations for A_i and B_j are solved iteratively, and it can easily be checked that they ensure that the T_{ij} given in Equation (2.5) satisfies the constraint Equations (2.3) and (2.4). Note also that c_{ij} in such a model should be interpreted as a general measure of impedance between i and j, which may be measured as actual distance, as travel time, as cost, or more effectively as some weighted combination of such factors sometimes referred to as a 'generalised cost'. With this proviso, Equations (2.4)–(2.7) describe a gravity model which has been extensively used, and the discussion above has shown its heuristic derivation by analogy with Newton's gravitational law.

There are other possible approaches to elementary distribution models. Perhaps the most popular alternative to the gravity model is the intervening opportunities' model. This is presented separately in Appendix 3. Appendix 3 includes a maximum-entropy derivation of the model. Meanwhile, we proceed with the maximum-entropy derivation of the gravity model, initially using a different physical analogy.

The principle of the method for this particular case has already been exhibited in Chapter 1. It was shown to be necessary to introduce a constraint equation on T_{ij} in addition to supplement Equations (2.3) and (2.4). This equation is

$$\sum_i \sum_j T_{ij} c_{ij} = C \ . \tag{2.8}$$

The most probable distribution is then the matrix $\{T_{ij}\}$ which maximises the entropy

$$W(\{T_{ij}\}) = \frac{T!}{\prod_{ij} T_{ij}!} \tag{2.9}$$

(where T is the total number of trips for this purpose) subject to these constraints, Equations (2.3), (2.4), and (2.8). We now proceed to the derivation, which was not given in Chapter 1. Recall that $W(\{T_{ij}\})$ is the total number of system *states* giving rise to the distribution $\{T_{ij}\}$. So the total number of possible states is w, where

$$w = \sum W(\{T_{ij}\}) \tag{2.10}$$

and the summation is over T_{ij}, satisfying Equations (2.3), (2.4), and (2.8). However, the maximum value of $W(T_{ij})$ turns out to dominate the other terms of the sum to such an extent that the distribution $\{T_{ij}\}$, which gives rise to this maximum, is overwhelmingly the most probable distribution. This maximum will now be obtained, and its sharpness, and the validity of the method in general, will then be discussed below.

To obtain the set of T_{ij}'s which maximises $W(\{T_{ij}\})$ as defined in Equation (2.9) subject to the constraints (2.3), (2.4), and (2.8), the

Lagrangian $\mathcal{L}$ has to be maximised, where

$$\mathcal{L} = \ln W + \sum_i \lambda_i^{(1)}\left(O_i - \sum_j T_{ij}\right) + \sum_j \lambda_j^{(2)}\left(D_j - \sum_i T_{ij}\right) + \beta\left(C - \sum_i \sum_j T_{ij} c_{ij}\right) , \tag{2.11}$$

and where $\lambda_i^{(1)}$, $\lambda_j^{(2)}$, and β are Lagrangian multipliers. Note that it is more convenient to maximise $\ln W$ rather than W, and then it is possible to use Stirling's approximation

$$\ln N! = N \ln N - N \tag{2.12}$$

to estimate the factorial terms. The T_{ij}'s which maximise $\mathcal{L}$, and which therefore constitute the most probable distribution of trips, are the solutions of

$$\frac{\partial \mathcal{L}}{\partial T_{ij}} = 0 \tag{2.13}$$

and the constraint Equations (2.3), (2.4), and (2.8). Using Stirling's approximation, (2.12), note that

$$\frac{\partial \ln N!}{\partial N} = \ln N \tag{2.14}$$

and so

$$\frac{\partial \mathcal{L}}{\partial T_{ij}} = -\ln T_{ij} - \lambda_i^{(1)} - \lambda_j^{(2)} - \beta c_{ij} \tag{2.15}$$

and this vanishes when

$$T_{ij} = \exp(-\lambda_i^{(1)} - \lambda_j^{(2)} - \beta c_{ij}) . \tag{2.16}$$

Substitute in Equations (2.3) and (2.4) to obtain $\lambda_i^{(1)}$ and $\lambda_j^{(2)}$:

$$\exp(-\lambda_i^{(1)}) = O_i \left[\sum_j \exp(-\lambda_j^{(2)} - \beta c_{ij})\right]^{-1} \tag{2.17}$$

$$\exp(-\lambda_j^{(2)}) = D_j \left[\sum_i \exp(-\lambda_i^{(1)} - \beta c_{ij})\right]^{-1} . \tag{2.18}$$

To obtain the final result in more familiar form, write

$$A_i = \frac{\exp(-\lambda_i^{(1)})}{O_i} \tag{2.19}$$

and

$$B_j = \frac{\exp(-\lambda_j^{(2)})}{D_j} \tag{2.20}$$

and then

$$T_{ij} = A_i B_j O_i D_j \exp(-\beta c_{ij}) , \tag{2.21}$$

where, using Equations (2.17)–(2.20),

$$A_i = \left[\sum_j B_j D_j \exp(-\beta c_{ij})\right]^{-1} \tag{2.22}$$

$$B_j = \left[\sum_i A_i O_i \exp(-\beta c_{ij})\right]^{-1} . \tag{2.23}$$

Thus the most probable distribution of trips is the same as the gravity model distribution discussed earlier, and defined in Equations (1.6)–(1.8) of Chapter 1 and Equations (2.5)–(2.7) above, with the negative exponential function replacing f. This statistical derivation constitutes a new theoretical base for the gravity model. Note that C in the cost constraint Equation (2.8) need not actually be known, as this equation is not in practice solved for β. This parameter would be found by the normal calibration methods. However, if C were known, then Equation (2.8) could be solved numerically for β.

This statistical theory is effectively saying that, given total numbers of trip origins and destinations for each zone for a homogeneous person-trip-purpose category, given the costs of travelling between each zone, and given that there is some fixed total expenditure on transport in the region, then there is a most probable distribution of trips between zones, and this distribution is the same as the one normally described as the gravity model distribution. Students of statistical mechanics will recognise the method as a variation of the microcanonical ensemble method for analysing systems of particles, for example the molecules of a gas.

2.2.2 Interpretation of terms

It has always been a feature of statistical mechanics that the terms which occur in the equation giving the most probable distribution are then seen to have physical significance. This is true here also. O_i, D_j, and c_{ij} were defined previously. The expression $\exp(-\beta c_{ij})$ appears in this formulation as the preferred form of distance deterrence function, and the parameter β is determined in theory by the cost-constraint Equation (2.8). It does, however, have its usual interpretation: it is closely related to the average distance travelled. The greater β, the less is the average distance travelled. This is obviously related to C of Equation (2.8). If C is increased, then more is spent on travelling and distances will increase, and an examination of the left-hand side of Equation (2.8) shows that β would decrease.

The remaining task is to interpret A_i and B_j. Suppose one of the D_j's changes, say, D_1. Then

$$T_{i1} = A_i B_1 O_i D_1 \exp(-\beta c_{i1}) \tag{2.24}$$

and, if D_1 changes substantially, the trips from each i to zone 1 would change in proportion. The next largest change will be in the A_i's as defined by Equation (2.22), but the change will not be large, as the expression involving D_1 in each A_i is only one of a number of terms. The B_j's will probably be affected even less, as any change is brought about through changes in the A_i's.

Suppose, then, that D_1 is substantially increased. Then the T_{i1}'s will increase more or less in proportion. The A_i's will decrease by a lesser amount relatively, and the B_j's will increase even more slightly. The role of the A_i's, then, will be to reduce all trips slightly to compensate for the increase in trips to zone 1. A_i can thus be seen as a competition term which reduces most trips owing to the increased attractiveness of one zone. The denominator of A_i is also commonly used as a measure of accessibility, and the increase in D_1 could be said to increase the accessibility of everyone to opportunities at 1, though more usually such an interpretation would be reserved for changes in the c_{ij}'s. Thus this analysis establishes a competition-accessibility interpretation of the A_i's. The B_j's play a similar role, and would be responsible for the main adjustments if the major change were in an O_i rather than in a D_j. A change in the c_{ij}'s, or several O_i's and D_j's simultaneously, would bring about complex readjustments through the A_i's and B_j's.

One consequence of this interpretation, and of the use of the new method which gives a fundamental role to the A_i's and B_j's, is that it shows that the interpretation of the A_i's and B_j's suggested by Dieter (1962) is wrong. Dieter suggested that the A_i's and B_j's should be associated with terminal costs, say a_i and b_j in origin and destination zones i and j, respectively. This can be checked by replacing c_{ij} by $a_i + b_j + c_{ij}$ in the preceding analysis, and this gives for T_{ij}:

$$T_{ij} = A_i B_j O_i D_j \exp(-\beta a_i - \beta b_j - \beta c_{ij}) . \tag{2.25}$$

Thus new terms $\exp(-\beta a_i)$ and $\exp(-\beta b_j)$ are introduced, but the A_i's and B_j's are still present independently of the existence of terminal costs.

2.2.3 Validity of the method

There are two possible points of weakness in the method. Firstly, is Stirling's approximation in Equation (2.14) valid for the sort of T_{ij}'s that occur in practice? Secondly, is the maximum value of

$$\frac{T!}{\prod_{ij} T_{ij}!}$$

a very sharp maximum?

The first of the doubts can be answered by analogy. The use of Stirling's approximation underlies one particular approach in statistical mechanics and is used, as here, to produce most probable distributions. There is, however, a second method, the Darwin-Fowler method, which actually calculates the individual terms of the sum in Equation (2.10) by using a generating function and complex integration. These terms are then used as weights to calculate the means of all the distributions, and these mean values have been shown to be the same as the most probable values obtained by using Stirling's approximation, even in the cases where the numbers involved are obviously so small that Stirling's theorem is not valid. This alternative derivation is presented in Appendix 4. More fundamentally, the derivation is based only on the derivative of Stirling's approximation as noted earlier, and we will see later that this leads to the same derivative of entropy defined in Equation (2.9) as that arising from the information theorists' definition—and hence to the same results. The derivation via information theory does not depend on Stirling's approximation. The information theory derivation, using some of the results of later chapters, is given in Chapter 7. The definition of Equation (2.9) is, however, convenient in practice, as its derivative is slightly more convenient to handle.

The second question can be answered explicitly if small changes in $\ln w(T_{ij})$ are examined near the maximum. At, or very near, the maximum the terms of $\mathrm{d}[\ln w(T_{ij})]$ which are linear in $\mathrm{d}T_{ij}$ vanish, and

$$\mathrm{d}[\ln w(T_{ij})] = \tfrac{1}{2}\sum_i\sum_j \frac{\partial^2 \ln w}{\partial T_{ij}^2}(\mathrm{d}T_{ij})^2 \ . \tag{2.26}$$

It will be recalled that

$$\frac{\partial \ln w}{\partial T_{ij}} = -\ln T_{ij} \ , \tag{2.27}$$

and so

$$\frac{\partial^2 \ln w}{\partial T_{ij}^2} = -T_{ij}^{-1} \ . \tag{2.28}$$

Substituting in Equation (2.26), we have

$$\mathrm{d}[\ln w(T_{ij})] = -\tfrac{1}{2}\sum_i\sum_j \frac{(\mathrm{d}T_{ij})^2}{T_{ij}} = -\tfrac{1}{2}\sum_i\sum_j \left(\frac{\mathrm{d}T_{ij}}{T_{ij}}\right)^2 T_{ij} \ . \tag{2.29}$$

Thus Equation (2.29) can be written

$$\mathrm{d}[\ln w(T_{ij})] = -\tfrac{1}{2}\sum_i\sum_j p^2 T_{ij} \tag{2.30}$$

where p is the percentage change in each T_{ij} away from the most probable distribution.

To evaluate this expression, the size distribution of elements of the trip matrix is needed. Suppose there are N size groups, and that the nth group has T_{ij}'s with a mean value T_n, and there are S_n such trip matrix elements in this group. Then Equation (2.30) can be written

$$\mathrm{d}[\ln w(T_{ij})] = -\tfrac{1}{2}p^2 \sum_n S_n T_n \ . \tag{2.31}$$

Now consider a typical example: take a large urban area with, say, 1000 zones. Suppose 1000 trip interchanges have 10^4 trips each, 10000 have 10^3, and 100000 have 10^2. Let $p = 10^{-3}$. Then

$$d\ln w \approx -\tfrac{1}{2}10^{-6}(10^7+10^7+10^7) \approx -15 .$$

Thus $\ln w$ changes by -15 for a change of one part in a thousand of each element of the trip matrix away from the most probable distribution. Thus w drops by the enormous factor of e^{-15}, which gives an indication of just how sharp the maximum can be. Such an estimate of $d(\ln w)$ can be calculated as a check in any particular case. One of the advantages of this new approach is that it gives the possibility of doing this check, and ruling out certain situations as being unsuitable for the gravity-model approach should the maximum turn out not to be a sharp one.

A third result of interest can also be stated by analogy with the corresponding result in statistical mechanics (cf. Tolman, 1938). That is,

$$\frac{\overline{(T_{ij}-\overline{T}_{ij})^2}}{\overline{T}_{ij}^2} = \overline{T}_{ij}^{-1} - T^{-1} . \tag{2.32}$$

This gives the dispersion of T_{ij} and indicates, as is well known in practice, that estimates are better for large flows than for small ones.

This analysis has shown that the gravity model has a sound base. However, it should be recalled that the whole analysis has been for a single trip purpose, and for a homogeneous troup of travellers. People are not identical in the way that particles in physics are identical, and so no theory of this form (indeed, no theory period) can be expected to apply exactly. This analysis has shown, in effect, that good results can be expected if trips can be classified by purpose and by person type in a reasonably uniform way. We now have to see how to construct generalised models to represent more complex situations.

2.3 Generalised distribution models

It is now possible to make the model more realistic by introducing several person types and several transport modes. The person types to be identified here are those which have different sets of modes available to them. For example, car owners (which is shorthand for the set of people who have cars available to them) have access to both car and public transport modes, while non car owners only have access to public transport modes. This categorisation at least seems essential. It may also be useful to bear in mind the possibility of dealing with people in different income groups or social classes in this way. If the car owner/non car owner division is not made, then this is likely either to give rise to models which allocate some non car owners to car trips, or it forces us to distribute car owner and non car owner trips separately and hence to have to forecast trip attractions for each group separately, and this is usually neither possible nor desirable.

There is a need then, to extend our notation: let T_{ij}^{kn} denote the number of trips between i and j by mode k and person type n; let O_i^n be the number of trip origins at i generated by persons of type n; let c_{ij}^k be the cost of travelling from i to j by mode k; and define the other variables as before. Let $M(n)$ be the set of modes available to type n people, and let $k \in M(n)$ denote one such mode; $\sum_{k \in M(n)}$ denotes summation over such modes. Note that we should really define a quantity $M_{ij}(n)$, as the modes available between i and j to type n people, but we shall allow the i and j subscripts to be understood on the assumption that usually all modes will be available between all pairs of zones. The subscripts can easily be reintroduced if necessary.

A further development of notation is that when an index is replaced by an asterisk, then this denotes summation over that index. For example,

$$T_{ij}^{*n} = \sum_{k \in M(n)} T_{ij}^{kn} \tag{2.33}$$

is the total number of trips by n-type people between i and j (that is, T_{ij}^{kn} summed over modes k).

We can now begin to write down constraint equations which express our hypotheses about the new situation. Firstly, we must make a hypothesis about trip generation. It is customary (see again, for example, Wootton and Pick, 1967) to characterise home-based trip ends (productions) by person type and non-home-based trip ends (attractions) by another zone characteristic such as land use (activity). To put this another way, we might expect O_i to be a function of n, but not D_j, as we make the assumption that different types of people, within a trip purpose, compete for the same attractions. This is why trip origins have been defined above as O_i^n and trip destinations as D_j. This seems useful and sensible, even when our only n categorisation is to car owners and non car owners. Then

$$\sum_j \sum_{k \in M(n)} T_{ij}^{kn} = O_i^n \tag{2.34}$$

$$\sum_i \sum_n \sum_{k \in M(n)} T_{ij}^{kn} = D_j \tag{2.35}$$

describe our trip end hypotheses. We might also hypothesise that different person types have different per capita expenditure on travel, and hence different travel patterns, and so we write a cost constraint for each person type n as

$$\sum_i \sum_j \sum_{k \in M(n)} T_{ij}^{kn} c_{ij}^k = C^n . \tag{2.36}$$

Now maximise

$$\frac{\ln T!}{\prod_{\substack{ijn \\ k \in M(n)}} T_{ij}^{kn}!}$$

subject to Equations (2.34)–(2.36) as constraints in the usual way, and we obtain

$$T_{ij}^{kn} = A_i^n B_j O_i^n D_j \exp(-\beta^n c_{ij}^k) , \tag{2.37}$$

where

$$A_i^n = \left[\sum_j \sum_{k \in M(n)} B_j D_j \exp(-\beta^n c_{ij}^k)\right]^{-1} \tag{2.38}$$

and

$$B_j = \left[\sum_i \sum_n \sum_{k \in M(n)} A_i^n O_i^n \exp(-\beta^n c_{ij}^k)\right]^{-1} . \tag{2.39}$$

Note that Equation (2.37) is a linked set of gravity models for each k–n category. The linking is through the B_j's, which each involve all the k's and n's. It is possible to program this model for a computer, and such a model is in use at the Ministry of Transport in London (see Wilson, Hawkins, Hill, and Wagon, 1969). There are two ways in which this generalised distribution model can be aggregated: over k and over n.

It is felt, for example, that one person type is sufficient to categorise a population (for example, when nearly everyone has a car available); then we can aggregate over n to obtain this single category, and[(3)]

$$T_{ij}^{k*} = A_i^* B_j O_i^* D_j \exp(-\beta^* c_{ij}^k) \tag{2.40}$$

$$A_i^* = \left[\sum_j \sum_k B_j D_j \exp(-\beta^* c_{ij}^k)\right]^{-1} \tag{2.41}$$

$$B_j = \left[\sum_i \sum_k A_i^* O_i^* \exp(-\beta^* c_{ij}^k)\right]^{-1} \tag{2.42}$$

is the appropriate model. Note that, if this model is in turn aggregated over k, then we obtain the original gravity model represented in Equations (2.21)–(2.23). It remains useful, however, to aggregate the general model, Equations (2.37)–(2.39), over k to represent a single-mode situation, or simply a model for trip interchanges by person type:

$$T_{ij}^{*n} = A_i^n B_j O_i^n D_j \sum_{k \in M(n)} \exp(-\beta^n c_{ij}^k) \tag{2.43}$$

$$A_i^n = \left[\sum_j B_j D_j \sum_{k \in M(n)} \exp(-\beta^n c_{ij}^k)\right]^{-1} \tag{2.44}$$

$$B_j = \left[\sum_i \sum_n A_i^n O_i^n \sum_{k \in M(n)} \exp(-\beta^n c_{ij}^k)\right]^{-1} . \tag{2.45}$$

Notice that each k summation is over the modes available to persons of type n.

(3) In these equations, of course, the asterisk associated with the A_i^* and β^* does not indicate any summation, but merely that new parameters are involved.

It is useful to derive one more model in this section. Suppose that, instead of having travel costs presented as an array c_{ij}^{k}, we have been given C_{ij}^{n}, the cost of getting from i to j as perceived by an n-type person. C_{ij}^{n} is clearly a composite of the c_{ij}^{k}'s for the k's available to each person type n, and we shall return to this later. Then, using the variable T_{ij}^{n} and maximising $\ln T! \Big/ \prod_{ijn} T_{ij}^{*n}!$ subject to

$$\sum_{j} T_{ij}^{*n} = O_{i}^{n} \tag{2.46}$$

$$\sum_{i}\sum_{n} T_{ij}^{*n} = D_{j} \tag{2.47}$$

$$\sum_{i}\sum_{j} T_{ij}^{*n} C_{ij}^{n} = C^{n} \ , \tag{2.48}$$

we obtain

$$T_{ij}^{*n} = A_{i}^{n} B_{j} O_{i}^{n} D_{j} \exp(-\beta^{n} C_{ij}^{n}) \ , \tag{2.49}$$

where

$$A_{i}^{n} = \left[\sum_{j} B_{j} D_{j} \exp(-\beta^{n} C_{ij}^{n})\right]^{-1} \tag{2.50}$$

and

$$B_{j} = \left[\sum_{i}\sum_{n} A_{i}^{n} O_{i}^{n} \exp(-\beta^{n} C_{ij}^{n})\right]^{-1} \ . \tag{2.51}$$

To summarise: we have derived a generalised model represented by Equations (2.37)–(2.39), two aggregated versions represented by Equations (2.40)–(2.42) and (2.43)–(2.45), respectively, and one alternative version of Equations (2.43)–(2.45) computed directly and represented by Equations (2.49)–(2.51). The implications of these models will be fully discussed in the following sections.

2.4 Modal-split implications

Firstly, consider the model given by Equations (2.40)–(2.42), which emphasises modes only. Sum Equation (2.40) over k, and divide the result into (2.40) itself to give the *modal split* as

$$M_{ij}^{k} = \frac{T_{ij}^{k*}}{T_{ij}^{**}} = \frac{\exp(-\beta^{*} c_{ij}^{k})}{\sum_{k'} \exp(-\beta^{*} c_{ij}^{k'})} \ . \tag{2.52}$$

This can be written

$$M_{ij}^{k} = \frac{1}{1 + \sum_{k' \neq k} \exp[-\beta^{*}(c_{ij}^{k'} - c_{ij}^{k})]} \tag{2.53}$$

This is a modal-split model which is implicit in our generalised distribution model. The Equation (2.53), for two modes, represents something rather like a diversion curve, but emphasises cost differences rather than the more usual cost ratios. It can also deal with any number of modes. Further, it is of considerable interest that Equation (2.52) has the same form as the modal-split equation derived from an application of discriminant analysis (cf. Quarmby, 1967) provided that the generalised cost function can be identified with the statistician's discriminant function. This apparent identity suggests that discriminant analysis is a good way of estimating the weights in a generalised cost function of the form

$$c_{ij}^{k} = \sum_{r} a_r X_r(i, j, k) \tag{2.54}$$

where the X_r's are variables like fares, travel time, and excess time, as mentioned earlier.

We can now examine the modal split implications of the general model, Equations (2.37)–(2.39). By a similar procedure to the above, we can obtain a modal split for each person type in the form

$$M_{ij}^{kn} = \frac{T_{ij}^{kn}}{T_{ij}^{*n}} = \frac{\exp(-\beta^n c_{ij}^k)}{\sum_{k \in M(n)} \exp(-\beta^n c_{ij}^k)} \tag{2.55}$$

but where, now, the mode summation is over the subset of modes available to persons of type n. The assumptions underlying these models, which are combined distribution and modal-split models, will now be made more explicit and discussed in the next section.

2.5 Review of underlying behavioral hypotheses

Firstly, note that C^n, the expenditure on this class of trips by n-type people, must exceed some minimum amount in order for a T_{ij}^{kn} to exist which meets the constraints relating the trip matrix to known numbers of trip ends. In fact, as

$$\beta^n \to \infty, \qquad \sum_{ijk} T_{ij}^{kn} c_{ij}^k \to \text{a minimum,}$$

where this minimum is the minimum value of C^n. Insofar as C^n exceeds this minimum, as demonstrated by calibrated values of β^n, the excess represents what people are prepared to pay, and what people actually do pay, partly to travel further than the minimum distances, and partly to travel by more expensive modes.

In a logistic modal-split formula, of the form Equation (2.55), β^n measures the sensitivity of type n people to the costs of different modes. If β^n is small, then there is little price discrimination between modes, but if it is large, the majority of people travel by the minimum-cost mode. A large β^n corresponds to a small C^n, of course.

However, β^n also plays another role in the distribution models, say in Equations (2.37)–(2.39), for example. It determines the average length of trip, and so also the sensitivity of people to trip length. If β^n is small, average trip lengths are long (and C^n is large) and *vice versa*. The main question to raise is whether a single parameter, β^n, for each person type, should do the same job for mode choice and what might be called trip-length choice.

A second major point is related to this discussion. The model represented by Equations (2.49)–(2.51) was derived on the assumption that we were given C_{ij}^n, the cost of travelling from i to j as perceived by an n-type person, and it was remarked that this must be some composite of the modal costs c_{ij}^k. This question of determining the form of composite impedance or cost, which is a question of long standing in the construction of transport-demand models, turns out to be very important in an analysis of alternative modal-split models. Note that the model represented by Equations (2.49)–(2.51) can be wholly identified with the aggregate generalised model represented by Equations (2.43)–(2.45) if

$$\exp(-\beta'^n C_{ij}^n) = X \sum_{k \in M(n)} \exp(-\beta^n c_{ij}^k) \; , \tag{2.56}$$

where β'^n has been written instead of β^n in the expression of the left-hand side of the equation, as the β^n's in the two sets of equations may not be the same. X is an arbitrary multiplicative constant. This then suggests itself as a good functional form for composite impedance, for determining C_{ij}^n as a function of the c_{ij}^k's.

Since C_{ij}^n is constructed out of the c_{ij}^k's, it is an array of quantities whose units are, in fact, determined by the functional relationship between the two arrays. In fact, it can easily be seen that, in Equation (2.56), X determines the position of the zero of the scale, and β^n determines the size of the unit on the scale. Generally, it will be convenient to take X as the reciprocal of the number of modes available (for all travellers), since this ensures that C_{ij}^n will always be positive, and to take β'^n equal to β^n, which will give the C_{ij}^n's the same scale of units as the c_{ij}^k's. We shall see shortly that there are alternative ways of expressing composite impedance relationships.

The earlier discussion of the roles of β^n implies that alternative modal-split hypotheses are possible. The next step is to discuss these alternatives.

2.6 An alternative modal-split model

The obvious alternative is to seek to set up a model structure which has one set of negative exponential coefficients β^n to determine average trip lengths, and another to determine modal split. We can then test empirically which set of hypotheses best fits the facts.

An alternative model can be set up using the entropy-maximising method as follows. Suppose a set of C_{ij}^n's exists, which are the generalised costs of travel between i and j as perceived by n-type people. Assume that these are the costs, however they are derived, which are relevant to the distribution of trips, and hence to trip lengths. In the earlier model, an amount C^n was spent by n-type people on these trips. Suppose now that some part of this, $\Gamma_1^{(n)}$, is spent on achieving destinations (that is, trip distribution), so that

$$\sum_i \sum_j T_{ij}^{*n} C_{ij}^n = \Gamma_1^{(n)} \tag{2.57}$$

$$\sum_j T_{ij}^{*n} = O_i^n \tag{2.58}$$

$$\sum_i \sum_n T_{ij}^{*n} = D_j \tag{2.59}$$

in the usual way, which gives as a maximum-entropy model

$$T_{ij}^{*n} = A_i^n B_j O_i^n D_j \exp(-\beta^n C_{ij}^n) , \tag{2.60}$$

in which A_i^n and B_j are given as in Equations (2.50) and (2.51). Suppose now that the balance of the amount C^n, $\Gamma_2^{(n)} = C^{(n)} - \Gamma_1^{(n)}$, is devoted to travel by other than minimum-cost modes between each pair of zones. Let n_{ij} represent the minimum cost between i and j, and then we can write

$$\sum_{k \in M(n)} (c_{ij}^k - n_{ij}) T_{ij}^{kn} = \Gamma_2^{(n)}(i, j) \tag{2.61}$$

where $\Gamma_2^{(n)}(i, j)$ is the $(i\text{-}j)$ part of $\Gamma_2^{(n)}$, and

$$\sum_{k \in M(n)} T_{ij}^{kn} = T_{ij}^{*n} . \tag{2.62}$$

Then, maximise

$$\frac{\ln T_{ij}^{*n}!}{\prod_{k \in M(n)} T_{ij}^{kn}!}$$

subject to Equations (2.61) and (2.62) in the usual way, to give

$$\frac{T_{ij}^{kn}}{T_{ij}^{*n}} = \frac{\exp[-\lambda_{ij}^n (c_{ij}^k - n_{ij})]}{\sum_{k \in M(n)} \exp[-\lambda_{ij}^n (c_{ij}^k - n_{ij})]} \tag{2.63}$$

which reduces to

$$\frac{T_{ij}^{kn}}{T_{ij}^{*n}} = \frac{\exp(-\lambda^n c_{ij}^k)}{\sum_{k \in M(n)} \exp(-\lambda^n c_{ij}^k)} \tag{2.64}$$

if we assume that the multiplier λ^n is independent of i and j. The n_{ij}'s cancel out. Thus we now have a distribution modal-split model which distributes trips by person type in the same way as the aggregated version, Equations (2.49)–(2.51), of the general model, Equations (2.37)–(2.39), and which has a modal split, Equation (2.64), of the same form as Equation (2.55), but with a different set of negative exponential coefficients. These alternative models could, in principle, be tested against survey data to see which gives the best fit.

2.7 Alternative estimates of composite impedance

There is a further degree of freedom, hinted at in Section 2.5, which we must not overlook: the construction of the array C_{ij}^n as a composite impedance out of the modal costs c_{ij}^k. In Section 2.5 it was shown that

$$\exp(-\beta'^n C_{ij}^n) = X \sum_{k \in M(n)} \exp(-\beta^n c_{ij}^k) \tag{2.56}$$

is one way of obtaining the C's as a function of the c's. We now see that the generalised model, Equations (2.37)–(2.39), and the corresponding modal-split Equation (2.55) imply that a set of C_{ij}^n's can be defined so that an aggregate model of the form of Equations (2.43)–(2.45) holds, and that these C_{ij}^n's are related to the c_{ij}^k's by Equation (2.56) with $\beta'^n = \beta^n$. However, we can now consider Equation (2.64) to be our most general modal-split formula, and so we now also see that, if Equation (2.56) holds as a composite impedance rule, then

$$\beta^n = \lambda^n , \tag{2.65}$$

and where we usually take

$$X = \frac{1}{N} , \tag{2.66}$$

where N is the total number of modes.

However, since we now admit that the modal-split mechanism can be separate from the distribution model, we also implicitly admit that composite impedance functions may be different. Thus, instead of constructing C_{ij}^n's from Equation (2.56), people may behave for trip distribution as though they perceived the minimum cost only of all the modes available to them. Thus, for distribution, we might have

$$C_{ij}^n = \min_{k \in M(n)} (c_{ij}^k) , \tag{2.67}$$

with Equations (2.60) and (2.64). This may give good results with β^n and λ^n estimated separately, and perhaps not being equal. We could also think of more general forms of composite impedance: for example,

$$C_{ij}^n = \frac{\sum_{k \in M(n)} w_{ij}^k c_{ij}^k}{\sum_{k \in M(n)} w_{ij}^k} \tag{2.68}$$

or

$$\exp(-\beta^n C_{ij}^n) = \frac{\sum_{k \in M(n)} w_{ij}^k \exp(-\beta^n c_{ij}^k)}{\sum_{k \in M(n)} w_{ij}^k} \tag{2.69}$$

where the w_{ij}^k's are weights. As it happens, because of our separate modal-split mechanism, Equation (2.64) (which can be calibrated separately from, and in fact preceding, the distribution process), we already know the proportions of travellers of each type on each trip interchange travelling by mode k, and this enables us to work out the proportions of all trips between i and j made by mode k. Then we could take w_{ij}^k to be this proportion, so

$$w_{ij}^k = \frac{\sum_n T_{ij}^{kn}}{\sum_n T_{ij}^{*n}} \tag{2.70}$$

which can be written as

$$w_{ij}^k = \frac{\sum_n T_{ij}^{kn}}{T_{ij}^{*n}} \frac{T_{ij}^{*n}}{\sum_n T_{ij}^{*n}} \;. \tag{2.71}$$

Then, substituting from Equations (2.60) and (2.64) for T_{ij}^{*n} and T_{ij}^{kn}/T_{ij}^{*n}, respectively, we have

$$w_{ij}^k = \frac{\sum_n \dfrac{\exp(-\lambda^n c_{ij}^k)}{\sum_{k \in M(n)} \exp(-\lambda^n c_{ij}^k)} A_i^n B_j O_i^n D_j \exp(-\beta^n C_{ij}^n)}{\sum_n A_i^n B_j O_i^n D_j \exp(-\beta^n C_{ij}^n)} \;. \tag{2.72}$$

This can then be substituted in Equations (2.68) and (2.69), and either equation could be solved iteratively for C_{ij}^n. Note that the solution takes a particularly simple form if $n = 1$ (and we are simply estimating a single composite cost for distribution from modal costs), as C_{ij}^n cancels from the right-hand side. We thus have at least four suggestions for functional forms of composite measures of impedance, given by Equations (2.56), (2.67), (2.68), and (2.69). For convenience, we shall call these the C1, C2, C3, and C4 forms respectively.

It is now worth summarising the alternative models which have now been assembled for testing:

(a) There is the generalised model, Equations (2.37)–(2.39), with its implied modal split, Equation (2.55). This has the same set of negative exponential coefficients for determining average trip lengths, and modal split. It implies a C1 form of composite impedance for C_{ij}^n, if this is defined. Thus, formally, this combined model can be seen as a combination of a person-type distribution model, as represented by Equations (2.49)–(2.51), with coefficients β^n, and a separate modal-split mechanism, Equation (2.64), but with $\lambda^n = \beta^n$, and C_{ij}^n defined and constructed out of the c_{ij}^k's by a C1 type of composite impedance formula.

(b) The modal-split model can be separate, so that λ^n is not necessarily equal to β^n (it is a matter of empirical test), and that C_{ij}^n is not necessarily constructed out of the c_{ij}^k's by a C1 type of impedance formula: there are the alternatives, such as C2, C3, and C4.

For convenience, call the modal-split mechanism, represented by Equation (2.64), M1, and the distribution model, represented by Equations (2.49)–(2.51), D1.

2.8 Route split

The problem has still greater depth and, unfortunately, complication. c_{ij}^k has been defined as the cost of travelling from i to j by mode k, and we constructed our other costs, the C_{ij}^n's, using composite impedance formulas, out of these. But, what is a mode, and what are modal costs? Costs are observed on actual *routes*, and mode k, between i and j, may consist of several routes. For example, if we have two modes, car and rail, then there may be several road routes and several rail routes between each zone pair. Thus we need to define γ_{ij}^r to be the cost of travelling on the rth route between i and j. *This is what can be observed.* A mode can now be defined as a set of routes. Let $R_{ij}(k)$ be the set of routes between i and j which we define to be mode k, let $r \in R_{ij}(k)$ be one such route, and let $\sum_{r \in R_{ij}(k)}$ denote summation over such routes. It is also useful to define $M_{ij}(n)$ with the subscripts i and j restored. Note that $\sum_{\substack{r \in R_{ij}(k) \\ k \in M_{ij}(n)}}$ denotes summation over all routes available to n-type people.

We can now define the concept of route split, to complement that of mode split, and we can investigate how to construct the c_{ij}^k's out of the γ_{ij}^r's, using, again, composite impedance relationships, and also the interrelationship of route split and mode split. The concept of route split is, of course, particularly relevant to the assignment part of a transport model. For example, various kinds of capacity-restraint procedure, those which generate several routes between each pair of points, need allocation, or route-split formulas, to allocate travel between alternative routes. So the work of this section has direct implications for transport models, as well as being important for its relation to modal-split concepts.

Let S_{ij}^{rn} be the number of trips between i and j by persons of type n on the rth route between i and j. Then note

$$S_{ij}^{*n} = T_{ij}^{*n} . \tag{2.73}$$

There are at least two possible mechanisms which might determine route split within the maximum entropy model methodology, assuming:

(a) That people perceive route costs directly, and that a route-split formula can be developed by analogy with Equation (2.64), but using a parameter μ^n to allow for the possibility of its being different from λ^n. Then

$$\frac{S_{ij}^{rn}}{S_{ij}^{*n}} = \frac{\exp(-\mu^n \gamma_{ij}^r)}{\sum_{\substack{r \in R_{ij}(k) \\ k \in M_{ij}(n)}} \exp(-\mu^n \gamma_{ij}^r)} \tag{2.74}$$

is an appropriate route-split equation.

(b) That people perceive mode costs directly, and that mode split is determined by Equation (2.64):

$$\frac{T_{ij}^{kn}}{T_{ij}^{*n}} = \frac{\exp(-\lambda^n c_{ij}^k)}{\sum_{k \in M(n)} \exp(-\lambda^n c_{ij}^k)} \quad . \tag{2.64}$$

Route split is then determined *within* modes thus:

$$\frac{S_{ij}^{rn}}{T_{ij}^{kn}} = \frac{\exp(-\mu^n \gamma_{ij}^r)}{\sum_{r \in R_{ij}(k)} \exp(-\mu^r \gamma_{ij}^r)} \quad . \tag{2.75}$$

Call these the R1 and R2 views of route split, for convenience. Once again, as mentioned earlier, we have a composite impedance problem. Consider the functional forms examined previously, and apply them to the problem of constructing c_{ij}^k's out of γ_{ij}^r's:

$$\text{C1:} \quad \exp(-\lambda^n c_{ij}^k) = X \sum_{r \in R_{ij}(k)} \exp(-\mu^n \gamma_{ij}^r) \tag{2.76}$$

$$\text{C2:} \quad c_{ij}^k = \min_{r \in R_{ij}(k)} (\gamma_{ij}^r) \tag{2.77}$$

$$\text{C3:} \quad c_{ij}^k = \frac{\sum_{r \in R_{ij}(k)} \gamma_{ij}^r w_{ij}^r}{\sum_{r \in R_{ij}(k)} w_{ij}^r} \tag{2.78}$$

$$\text{C4:} \quad \exp(-\lambda^n c_{ij}^k) = \frac{\sum_{r \in R_{ij}(k)} w_{ij}^r \exp(-\mu^n \gamma_{ij}^r)}{\sum_{r \in R_{ij}(k)} w_{ij}^r} \quad . \tag{2.79}$$

Note, firstly, that if C1 is the correct form, then we can take $\mu^n = \lambda^n$, and then the R1 and R2 mechanisms can hold simultaneously. In fact, that C1 should hold is a necessary and sufficient condition for this. To put this the other way round, if C2, C3, C4 or some functional form other than C1 emerges as the correct functional form for estimating modal impedances from route impedances, and, if a mode split model fits the facts, then the R1 view is untenable. If this turned out to be the case, this would be a strong statement about perception. It would mean that people perceived modes as entities in some sense prior to routes.

To summarise: if an R1 view were justified, then perception would be route dominated; if an R2 view were justified, then perception would be mode dominated unless the appropriate form of composite impedance was C1, in which case the perception of modes and routes would have equal 'strength'. There is one other way in which this could be true: it is just theoretically possible that a composite impedançe functional form exists which allows R1 to be satisfied, together with the modal-split part of R2 [that is, Equations (2.74) and (2.64), respectively]. In such a case, we could not really say that R2 is satisfied simultaneously with R1, since

Equation (2.75) is not satisfied, but that R1, Equation (2.74), is satisfied simultaneously with M1.

2.9 Concluding comments and discussion

2.9.1 Introduction

It is obvious that the various generalisations of the simple transport gravity model outlined in this chapter take us a considerable distance away from models commonly in use. As mentioned earlier, generalised distribution and modal-split models are now being used, though there will probably be a considerable time lapse before the alternative route-split models can be tested. This is the case on two counts: firstly, there are mathematical difficulties involved with finding more than the 'best' route between a pair of points (Sakarovitch, 1968); and, secondly, there is little or no available data on route split, in contrast to the situation with distribution and mode split.

However, it is hoped that the entropy-maximising method adds something to the existing stock of knowledge. In this concluding discussion, we shall focus on a number of distinct (often unconnected) points in addition to those already made, concentrating on what the entropy-maximising method offers, rather than on a general review of the problem of transport modelling.

2.9.2 A simple example

Some of the points which have been made may become clearer if a simple example is discussed. Suppose there are only two points A and B in our system, and there are two routes between them, say one road (car) and one rail. Suppose the cost of travel on each route is identical. The split of traffic between A and B, route or mode, will be 1 : 1. Suppose now an additional road route is introduced, again all costs being identical and the same as before. The situations are illustrated in Figure 2.1. What is the split in the new situation? The R1 model gives the route split as 1 : 1 : 1 (road, road, rail), and hence the mode split (by summing over appropriate routes, not by applying a formula since there is no formula in this case) as 2 : 1 (road, rail), and so there has been a shift from rail to road (assuming fixed total number of passengers, and neglecting any generated trips). However, if C2, C3, or C4 impedance is used, then the R2 mechanism gives the route split in the ratios 1 : 1 : 2 and the mode split as 1 : 1. The results given by applying C2, C3, and C4 impedance formulas are identical in this simple illustration, but would not be in a more general case. Thus

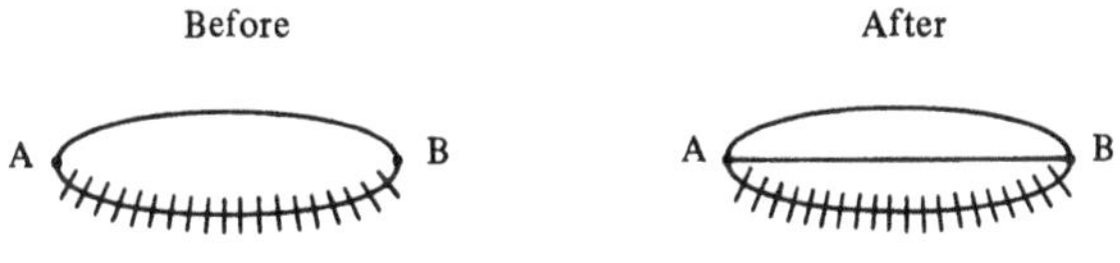

Figure 2.1. All costs equal, before and after.

quite different results can be obtained according to which model is chosen. It should be possible to seek out real-life situations which approximate this kind of example, and should allow us to discriminate between the models.

The example suggests that, *a priori*, R2 is preferable to R1, since the R1 model shifts some rail passengers to road with the introduction of a new route, even though no costs have changed. One might consider that the new road was an irrelevant alternative for marginal rail passengers, as they previously had the option of travelling by road at the newly available cost, and did not take it. If this feeling were borne out by empirical work, then this would throw a lot of stress on what we define to be a mode. For example, should we just have car and public transport or should express and local train services be counted as different modes; should toll motorways be counted as different from ordinary roads in this sense? A psychological motivation study may throw light on this. This problem was discussed by Quandt and Baumol (1966) in a paper on modal split, but not really answered.

The decision between R1 and R2 models would also have implications for capacity-restraint assignment algorithms which allocate traffic between alternative routes. What this work has shown is that (a) the algorithms should allocate between all routes in *joint* public–private networks if the R1 theory holds, (b) they should allocate between routes on *modal* networks only, with previously determined modal splits, if R2 holds, and (c) either can be used if, and only if, the C1 form of composite impedance relates modal costs and route costs.

2.9.3 Alternative impedance functions

It has been argued in this paper that the negative exponential function plays a major role in distribution and split models, and the success of the method advocated in this paper may seem to rest heavily on this particular function being successful in practice. However, there are several ways of changing the function while retaining the framework of the paper[(4)]:

(1) The costs with which we are dealing in the models are perceived costs, and there may be a transformation relating these to directly measured costs. This transformation could change the shape of the function, expressed as a function of directly measured costs.

(2) Trips could be classified into groups, such as long or short, each of which are considered to be more homogeneous than the complete set, and different functions applied to each.

(3) Harris (1964) has shown that if β^n, in this book's notation, has a gamma distribution in the population, then the negative exponential function is transformed into a power function.

It is worth discussing one of these as an example. In fact, we can use a

(4) I am indebted to Britton Harris for suggesting some of these.

transformation of the type suggested in (1) above to transform the negative exponential function into a power function in a simpler way than suggested in (3). Suppose people perceive travel costs not as we measure them, but as the logarithm of what we measure. Such an assumption would apply if the cost of travelling 50 miles was perceived to be less by the traveller who was committed to 200 miles anyway than to the traveller who was going 50 miles in total. Then c_{ij} is replaced by $\ln c_{ij}$, and $\exp(-\beta c_{ij})$ becomes $\exp(-\beta \ln c_{ij})$, which is $c_{ij}^{-\beta}$. Thus, if models fit better with inverse power functions than with negative exponential functions, this tells us something about the way travellers perceive costs. The rest of the machinery for transport modelling which has been developed still stands.

Thus we are not as committed to the negative exponential function as might appear at first sight. It will be used in general in the rest of the book, however, and the comments of this section should be borne in mind.

2.9.4 Alternative allocation formulas

There are also alternative allocation formulas that could be considered in addition to the possibility of changing the functional form of the generalised cost. If the generalised cost is given, as suggested, by Equation (2.54), then Quandt and Baumol (1966) use an allocation formula which is related directly to the X_r's:

$$\frac{T_{ij}^{k*}}{T_{ij}^{**}} = \frac{X_1(ijk)^{\alpha_1} X_2(ijk)^{\alpha_2} \ldots}{\sum_k X_1(ijk)^{\alpha_1} X_2(ijk)^{\alpha_2} \ldots} . \tag{2.80}$$

Other functions of the X_r's can also be written down. They each have different elasticity properties, and the representation of these properties, and comparison with observed behaviour, may be a good route into empirical testing. This is discussed in more detail by Hyman and Wilson (1969).

2.9.5 Procedure for estimating models

The methods to be used for estimating the models which have been outlined have not been discussed in detail at all, but one simple observation should be stated for modal-split or route-split models. Rather than use the form of Equation (2.64), it is better to rearrange the basic single mode formulas and represent the split as

$$\frac{T_{ij}^{k_1 n}}{T_{ij}^{k_2 n}} = \exp[-\lambda^n(c_{ij}^{k_1} - c_{ij}^{k_2})] \tag{2.81}$$

and then take logs.

2.9.6 A note on forecasting

When parameters of the form β^n have been estimated by some procedure which involves comparing model predictions with survey data, the model is then often used for forecasting within a planning context. This involves making some assumption about the behaviour of β^n over future time. It is customary to assume that β^n remains constant. The introduction of the constraint, Equation (2.36), repeated here for convenience:

$$\sum_i \sum_j \sum_{k \in M(n)} T_{ij}^{kn} c_{ij}^k = C^n \tag{2.36}$$

is an underlying hypothesis of the model. If β^n is estimated for the base year, then Equation (2.36) can be used to calculate C^n for the base year—the amount that n-type people spend on transport (measured of course in 'generalised' cost terms). In principle, it may be possible to estimate the rate of change of C^n from some independent spatially aggregated economic model. In such a case, Equation (2.36) could then be used to estimate the rate of change in β^n. In a recent paper (Hyman and Wilson, 1969), three possible hypotheses about the rate of change of β^n are investigated: that β^n is constant; that C^n is constant; and that $\beta^n C^n$ is constant. It is shown that the results obtained can vary quite sensitively with the hypothesis which is adopted, and so any advance on the lines outlined above would be a useful one.

3

Interregional commodity flows

3.1 Introduction

3.1.1 The problems to be studied

This chapter is concerned with a number of approaches to the analysis of interregional commodity flows. We first discuss a number of existing approaches, each covering aspects of the problem; a more than usually systematic appraisal of the Newtonian gravity model is given. The principles of the entropy-maximising method in relation to transport models are then taken up again from Chapter 2, and this enables a deeper study of gravity models of interregional flows to be made; the methods of this approach are then integrated into an interregional input–output framework.

It is interesting to note that this whole area of work falls between geography and economics: the development of the appropriate concepts of spatial interaction stem mainly from geography (helped by contributions from social physics!), and the development of the input–output framework stems mainly, of course, from economics. It will be shown that it is necessary to develop an integrated model. Further, the approaches to modelling spatial interaction, which will be developed in this chapter to interregional commodity flows, clearly have wider implications. Since relatively few models of interregional commodity flows have been tested, we are not constrained either by fashion, on the one hand, or by bounds determined by empirical testing, on the other, in exploring the form of models which might be useful. So this problem, being relatively unstudied, creates the freedom, for example, to study a wide variety of gravity models, and reveals that the gravity model must not be seen as a single model, but as a whole spectrum of possible models with subtle differences between one model and the next determined by shifts of basic assumptions. Thus, although commodity flows are interpreted in this chapter to mean the flows of goods, other interpretations are possible, enabling the models which are presented to be applied in other fields. Some of the results of this chapter will then be applied in later chapters.

3.1.2 Relevance to other problems

A model of interregional commodity flows would have a direct use for national planners: it would enable forecasts to be made of the impact of economic growth on the national freight transport system. Such a model would also be an aid to improving economic forecasts, particularly at the regional level, but in a way which also helps national and local planners. Regional economic forecasts are obviously of direct use to regional planners; they help national planners in relation to the policy issues associated with the allocation of resources between regions; they help local planners by facilitating the development of economic forecasts for cities, and by providing a regional context for local planning.

At present, urban or regional economic forecasting is still often carried out using the theory of the economic base. Artle (1961) has shown that the economic base method is outmoded and that a better understanding of an urban or regional economy can be achieved using the framework of the input–output model. The input–output model itself has deficiencies: on the one hand, we shall see that it has an extensive exogenous sector which must be forecast outside the model; on the other, cities and regions function in a wider interregional economic system. This chapter is particularly concerned with the second of these issues. If it is possible to find ways of modelling interregional commodity flows within an input–output framework, then it is a contribution to this second issue and a step towards better economic forecasting models.

3.1.3 Notation

Suppose we have a set of regions labelled by $i, j, k, \ldots$; let economic goods be classified into commodity groups labelled $m, n, p, \ldots$. The interregional transport network can be specified by the costs of transporting particular commodities between regions. Regions will be extensive spatially, and we define these costs to be the means of what would obviously be a range of costs. Further, for the present, we assume that there is only one mode of transport. Thus, let c_{ij}^m be the mean cost of carrying a unit of commodity m from an origin i to a destination j. Note that c_{ii}^m is the mean cost of carrying a unit of m from an origin in i to a destination also within i. Let x_{ij}^m be the total flow of commodity m from region i to region j measured in units appropriate to the commodity group.

If a subscript or superscript is replaced by an asterisk, this denotes summation over that index, as in Chapter 2. So, for example,

$$x_{i*}^m = \sum_j x_{ij}^m \tag{3.1}$$

is the total amount of commodity m produced in region i. Similarly, x_{*j}^m is the total amount of commodity m used in the production of other commodities or consumed by a final demand sector in j. It will also be convenient to define X_i^m and Y_i^m to be total production and consumption (now using 'consumption' to mean 'use' by all sectors, including final demand) in region i. This is useful in cases where these quantities are estimated by a procedure which is independent of that which estimates x_{ij}^m. It will also be useful to let X^m, without a regional subscript, represent the total production in the whole system, so

$$X^m = \sum_i X_i^m \tag{3.2}$$

and, of course, also equals $\sum_j Y_j^m$, as we are assuming the total system to be closed. (Closure can always be achieved by adding an additional region for 'the rest of the world'.)

Additional notation will be defined as the need arises.

3.2 Gravity models

3.2.1 The Newtonian gravity model, and some heuristic developments

3.2.1.1 *The Newtonian analogy*

Any derivation of the gravity model is customarily based on analogies between spatial interaction in geography and spatial interaction in classical physics, as we saw in the person-trip case in Chapter 2. Thus, in terms of the variables just defined, X_i^m and Y_j^m can be considered to be 'masses' of commodity m related to the origin and destination of a spatial interaction between regions i and j. The transport cost of a unit of commodity was defined to be c_{ij}^m and this can be considered to be a 'distance'. A strictly Newtonian interaction, still following the Chapter 2 argument by analogy, would be an x_{ij}^m defined by

$$x_{ij}^m = K^m \frac{X_i^m Y_j^m}{(c_{ij}^m)^2} , \tag{3.3}$$

where K^m is a normalising factor which ensures that

$$\sum_i \sum_j x_{ij}^m = X^m . \tag{3.4}$$

That is,

$$K^m = \frac{X^m}{\sum_i \sum_j X_i^m Y_j^m / (c_{ij}^m)^2} . \tag{3.5}$$

The first obvious development is to argue that geographic spatial interaction for commodity flows may well be governed by a distance function other than the inverse-square law, and Equation (3.3) could be amended to

$$x_{ij}^m = K^m X_i^m Y_j^m f^m(c_{ij}^m) , \tag{3.6}$$

where $f^m(c_{ij}^m)$ is some decreasing function of c_{ij}^m, and where K^m is now calculated from

$$K^m = \frac{X^m}{\sum_i \sum_j X_i^m Y_j^m f^m(c_{ij}^m)} . \tag{3.7}$$

Notice that we have allowed in our notation for the possibility of a different function for each commodity group. We shall see later that this is obviously necessary.

3.2.1.2 *The basic cases to be studied*

Further development is possible, but, as a preliminary, we must interpret our terms very carefully. Strictly, a model of interregional commodity flows provides estimates of x_{ij}^m, and hence, of course, of x_{i*}^m, x_{*j}^m, and x_{**}^m. However, x_{**}^m, and possibly x_{i*}^m and x_{*j}^m, may be estimated directly from independent models, and in our notation we have called such estimates X^m, X_i^m, and Y_j^m, respectively. There are four possible cases: firstly, there is an independent estimate of X^m, but not of X_i^m or Y_j^m; secondly, there is an independent estimate of X_i^m (which determines X^m), but not of Y_j^m; thirdly, there is an independent estimate of Y_j^m (which determines X^m), but not of X_i^m; fourthly, there are independent estimates of both X_i^m and Y_j^m (made in such a way that they determine X^m and that $\sum_i X_i^m = X^m$ and $\sum_j Y_j^m = X^m$). There are other conceivable cases; for example, where we have independent estimates of X_i^m and Y_j^m, but where the estimating procedure does not ensure that $\sum_i X_i^m = \sum_j Y_j^m$. But the four cases stated are the most interesting ones, and it will be clear how the methods of the paper could be applied to other cases if need be. Let these four cases be called case (i), case (ii), case (iii), and case (iv), respectively.

We can now carry out a further appraisal of the Newtonian form of the gravity model presented in Equation (3.6). Note, firstly, that in Equations (3.6) and (3.7), X_i^m should be replaced by x_{i*}^m, and Y_j^m by x_{*j}^m, in cases where they are not independently estimated. Since an estimate of X^m is assumed to exist in all cases, an equation of the form of Equation (3.7) can always be used to estimate K^m. Thus Equations (3.6) and (3.7), as they stand, represent the Newtonian gravity model for case (iv) and can easily be solved directly for x_{ij}^m. For each of cases (i)–(iii), the amended versions of Equations (3.6) and (3.7) lead to quadratic equations in x_{ij}^m which cannot easily be solved.

Note that in each of cases (i)–(iv) two of the following equations should be satisfied:

$$\sum_j x_{ij}^m = x_{i*}^m \tag{3.8}$$

$$\sum_i x_{ij}^m = x_{*j}^m \tag{3.9}$$

$$\sum_j x_{ij}^m = X_i^m \tag{3.10}$$

and

$$\sum_i x_{ij}^m = Y_j^m \ . \tag{3.11}$$

Equations (3.8) and (3.9) should hold for case (i), Equations (3.9) and (3.10) for case (ii), Equations (3.8) and (3.11) for case (iii), and Equations (3.10) and (3.11) for case (iv). In fact, it can easily be seen that the model given by Equations (3.6) and (3.7), and its variants, does not satisfy these obvious consistency checks in any case. There are two ways of viewing such a situation. Firstly, the degree of fit between each side of the appropriate equations in each case can be used as one of the measures by which the model is judged. Secondly, and alternatively, we can try to develop the model further to incorporate the equations as constraints.

This is the line which we now pursue.

Unfortunately, there are yet more possible variants of each development within what we have called cases (i)–(iii): these arise because either one of the two constraint equations for each case can be considered and applied as the only constraint, or, thirdly, both can be applied; the remaining equation of the first two of these variations could once again be used to help 'test' the resulting model—how well is it satisfied? In order to avoid writing out the algebra of each variant for each case, we make a simplifying assumption (without, as will be seen, loss of generality). That is, that only the equations out of Equations (3.8)–(3.11) with a variable on the right-hand side which is independently estimated [that is, Equations (3.10) and (3.11), in fact] will be used as constraints. Let us proceed on this assumption.

3.2.1.3 *Case* (i): *the unconstrained model*

Case (i) is unconstrained, given our simplifying assumption, and so Equations (3.6) and (3.7) continue to give the appropriate model under case (i) assumptions, though strictly x^m_{i*} and x^m_{*j} should replace X^m_i and Y^m_j, respectively.

3.2.1.4 *Case* (ii): *the production-constrained model*

Case (ii) has Equation (3.10) as a constraint on total production in region i. The use of the term 'production' also coincides with the use of this term in gravity models of passenger traffic flows. We can find a set of normalising factors to replace the single factor K^m which will ensure that Equation (3.10) is always satisfied. Define a set of factors A^m_i and modify Equation (3.6) to read

$$x^m_{ij} = A^m_i X^m_i x^m_{*j} f^m(c^m_{ij}) \tag{3.12}$$

[where x^m_{*j} has replaced Y^m_j in Equation (3.6) because it is not assumed to be independently estimated in case (ii)]. Then A^m_i can be calculated from Equation (3.10) if x^m_{ij} from Equation (3.12) is substituted in Equation (3.10), and this gives

$$A^m_i = \left[\sum_j x^m_{*j} f^m(c^m_{ij})\right]^{-1} , \tag{3.13}$$

and so Equations (3.12) and (3.13) represent the case (ii) model incorporating Equation (3.10) as a constraint. This is a heuristic derivation of a modified gravity model, and we shall produce this using entropy-maximising methods below.

3.2.1.5 *Case* (iii): *the attraction-constrained model*

For case (iii), it can easily be seen that Equation (3.11) is the constraint. This is called the attraction-constrained model, because the constraint operates on the total of each commodity attracted to a region, and this again brings it into line with terminology in passenger-transport gravity models. A set of normalising factors B_j^m replace K^m, and a similar calculation gives

$$x_{ij}^m = B_j^m x_{i*}^m Y_j^m f^m(c_{ij}^m) \tag{3.14}$$

where

$$B_j^m = \left[\sum_i x_{i*}^m f^m(c_{ij}^m)\right]^{-1} \tag{3.15}$$

3.2.1.6 *Case* (iv): *the production-attraction-constrained model*
Case (iv) is slightly different. Both of Equations (3.10) and (3.11) are constraints, and we need to replace K^m by two factors $A_i^m B_j^m$ to enable us to modify the model appropriately. Thus Equation (3.6) is modified to

$$x_{ij}^m = A_i^m B_j^m X_i^m Y_j^m f^m(c_{ij}^m) \ , \tag{3.16}$$

and the factors are calculated by substituting x_{ij}^m from Equation (3.16) into Equations (3.10) and (3.11), respectively, to give

$$A_i^m = \left[\sum_j B_j^m Y_j^m f^m(c_{ij}^m)\right]^{-1} \tag{3.17}$$

$$B_j^m = \left[\sum_i A_i^m X_i^m f^m(c_{ij}^m)\right]^{-1} \ , \tag{3.18}$$

and Equations (3.17) and (3.18) can easily be solved iteratively. Notice that, for a single commodity, the model given by Equations (3.16)–(3.18) has the same form as the person-trip model given by Equations (2.5)–(2.7) in Chapter 2, showing that the latter is what we have called a case (iv) (production-attraction-constrained) model.

3.2.1.7 *Discussion of the four cases*
Note that the case (iv) model is the only one which offers a simple estimate of x_{ij}^m; the other cases each give quadratic equations in the x_{ij}^m, although some iterative-solution procedure could be devised. We can also recover any apparent loss of generality incurred by our assumption about which constraint equations can be associated with each case. Consider case (ii), the production-constrained case, as an example. On our assumption, Equation (3.10) is the constraint. Two other possible variants were indicated: firstly, Equation (3.9) could be taken as a constraint, and Equation (3.10) not so taken; secondly, both of Equations (3.9) and (3.10) could be applied as constraints simultaneously. It can easily be seen that the first of these variants would in fact reduce to case (iii), the attraction-constrained case, with x_{*j}^m replacing Y_j^m, and the second reduces to case (iv), with x_{*j}^m replacing Y_j^m.

3.2.1.8 *Raising the Newtonian 'masses' to a power*
One further possible development must be mentioned and explored. Equation (3.6) contains the simple Newtonian hypothesis that x_{ij}^m is proportional to X_i^m and Y_j^m. It is possible to generalise this hypothesis

and suggest that x_{ij}^m should be proportional to X_i^m and Y_j^m, each raised to some power: say, $(X_i^m)^{\alpha^m}$ and $(Y_j^m)^{\beta^m}$, where α^m and β^m are additional parameters to be estimated. However, since this is the unconstrained case, we should, as remarked above, replace X_i^m by x_{i*}^m and Y_j^m by x_{*j}^m, so Equations (3.6) and (3.7) become

$$x_{ij}^m = K^m (x_{i*}^m)^{\alpha^m} (x_{*j}^m)^{\beta^m} f^m(c_{ij}^m) , \tag{3.19}$$

where

$$K^m = \frac{X^m}{\sum\sum (x_{i*}^m)^{\alpha^m} (x_{*j}^m)^{\beta^m} f^m(c_{ij}^m)} . \tag{3.20}$$

α^m and β^m could be estimated in this form of model by regression analysis. Note that if it was required that $K^m f^m$ was dimensionless, then the additional condition

$$\alpha^m + \beta^m = 1 \tag{3.21}$$

should be imposed in this particular form of model.

A similar development can be introduced into all other forms of gravity models so far discussed according to certain rules; all that is necessary is a statement of these rules rather than a detailed exposition. They are as follows: X_i^m (or x_{i*}^m if X_i^m is not independently estimated) can be replaced by $(X_i^m)^{\alpha^m}$ where it appears on the right-hand side of an equation, and Y_j^m (or x_{*j}^m if Y_j^m is not independently estimated) can be replaced by $(Y_j^m)^{\beta^m}$ where it appears on the right-hand side of an equation *provided that* X_i^m (or x_{i*}^m) or Y_j^m (or x_{*j}^m), as the case may be, do *not* appear in a constraint equation [that is, one of the set of Equations (3.8)–(3.11) in the model which is being extended].

Only the qualification needs to be explained. Take the case (ii) model given by Equations (3.12) and (3.13) as an example. According to the rules just stated, x_{*j}^m can be replaced by $(x_{*j}^m)^{\beta^m}$, but X_i^m *cannot* be replaced by $(X_i^m)^{\alpha^m}$ because it appears in constraint Equation (10) which is being recognised by this model. To show the need for the condition, suppose that X_i^m is replaced by $(X_i^m)^{\alpha^m}$ in Equations (3.12) and (3.13). Then, when x_{ij}^m from Equation (3.12) is substituted into Equation (3.10), having first substituted for A_i into Equation (3.12) from Equation (3.13), then Equation (3.10) reduces to $X_i^m = (X_i^m)^{\alpha^m}$, showing that α^m must be taken as 1. Hence the condition. Notice in relation to this digression that it is not possible to take a model of the form of Equation (3.19) and estimate α^m and β^m subject to the constraint Equations (3.10) and (3.11). Such a procedure would introduce factors A_i^m and B_j^m arising from the Lagrangian multipliers associated with Equations (3.10) and (3.11) and α^m and β^m would each be necessarily 1 as was the case for α^m in the above illustration. Thus, in practice, only terms of the form x_{i*}^m and x_{*j}^m (as distinct from X_i^m and Y_j^m) can be raised to a power.

3.2.1.9 *Summary*

The rather complex argument which has been put together in this systematic exploration of possible Newtonian gravity models is summarised in Table 3.1, which indicates how one set of models (for each of the four cases) develops out of alternative sets of assumptions.

Table 3.1. Summary of variants of Newtonian gravity model.

	Type of model	Case[5]	Equations in text	Constraints[6] recognised	Notes (or equation amendments)
1.	Simple Newtonian	(i)	(3.6), (3.7)		x_{i*}^m replaces X_i^m; x_{*j}^m replaces Y_j^m
2.		(ii)	(3.6), (3.7)		x_{*j}^m replaces Y_j^m
3.		(iii)	(3.6), (3.7)		x_{i*}^m replaces X_i^m
4.		(iv)	(3.6), (3.7)		
5.	Development to	(i)	(3.6), (3.7)		a number of variants
6.	take account of	(ii)	(3.12), (3.13)	(3.10)	are also discussed if
7.	constraints	(iii)	(3.14), (3.15)	(3.11)	alternative
8.	(constrained Newtonian)	(iv)	(3.16), (3.17) (3.18)	(3.10), (3.11)	combinations of constraints are imposed
	Generalisation which raises 'mass' terms to a power where possible, of:				
9.	(a) Simple Newtonian	(i)	(3.19), (3.20)		x_{i*}^m replaces X_i^m; x_{*j}^m replaces Y_j^m
10.		(ii)	(3.19), (3.20)		x_{*j}^m replaces Y_j^m
11.		(iii)	(3.19), (3.20)		x_{i*}^m replaces X_i^m
12.		(iv)	(3.19), (3.20)		
13.	(b) Constrained	(i)			rule given on p.42
14.	Newtonian	(ii)			for amending
15.		(iii)			models 5-8 above
16.		(iv)			

[5] Case (i): independent estimate of X^m only; case (ii): independent estimate of X_i^m; case (iii): independent estimate of Y_j^m; case (iv): independent estimate of X_i^m and Y_j^m.
[6] Any constraint which is not utilised (of the two possibles in each case) can be used to help 'test' the model.

3.2.2 The economists' gravity models

3.2.2.1 *Introduction*

There are at least two examples of gravity-type models being introduced directly into economic theories, and it is useful to explore these examples and contrast the models introduced to the discussion of Newtonian gravity models in the previous sections. The gravity models to be considered are those used by Theil (1967) and Leontief and Strout (1963).

3.2.2.2 *Theil's model*

Firstly, consider Theil's model, which is set up using extremely elementary assumptions, but which includes some interesting innovations. He suggests that the simplest conceivable model takes the form, in our notation,

$$X_{ij}^m = \frac{X_i^m Y_j^m}{X^m} . \tag{3.22}$$

He points out that it is unlikely that interregional commodity flows would satisfy such a simple relation (which is independent of distance—and he is considering international flows) and that a more realistic assumption might be

$$\frac{x_{ij}^m X^m}{X_i^m Y_j^m} = \frac{\hat{x}_{ij}^m \hat{X}^m}{\hat{X}_i^m \hat{Y}_j^m} , \tag{3.23}$$

where $\hat{x}_{ij}^m$, $\hat{X}^m$, $\hat{X}_i^m$, $\hat{Y}_j^m$ are known values of the variables at some base year. This can be written

$$x_{ij}^m = \frac{X_i^m Y_j^m}{X^m} Q_{ij}^m , \tag{3.24}$$

where

$$Q_{ij}^m = \frac{\hat{x}_{ij}^m \hat{X}^m}{\hat{X}_i^m \hat{Y}_j^m} , \tag{3.25}$$

and Q_{ij}^m can be considered to be an empirically estimated (and unsmoothed) 'distance' function.

3.2.2.3 *The gravity model of Leontief and Strout*

Theil's model has here been written in the form of Equation (3.24), because this is exactly the form postulated by Leontief and Strout, except that they do not assume that independent estimates of the regional totals X_i^m and Y_j^m are available. So the Leontief-Strout gravity model is

$$x_{ij}^m = \frac{x_{i*}^m x_{*j}^m}{x_{**}^m} Q_{ij}^m \tag{3.26}$$

and they assume that Q_{ij}^m can be estimated from base year data.

3.2.2.4 *Discussion: further developments by Theil*

Thus far, each of these models is a variant of what we have called the simple Newtonian gravity model, but with the distance function $f^m(c_{ij}^m)$ being replaced by an empirically determined factor Q_{ij}^m. Theil does, however, go further and he recognises some of the problems we have discussed in relation to the Newtonian model. Firstly, he notes that his estimate of x_{ij}^m in Equation (3.23) does not satisfy

$$\sum_i \sum_j x_{ij}^m = x_{**}^m , \tag{3.27}$$

and he suggests multiplication by a normalising factor to remedy this. He then notes that the estimate still does not satisfy Equations (3.10) and (3.11). He then suggests that the normalised estimate of x_{ij}^m should be replaced by $x_{ij}^{m'}$ obtained by minimising a quantity

$$I = \sum_i \sum_j x_{ij}^m \ln \frac{x_{ij}^m}{x_{ij}^{m'}} , \tag{3.28}$$

which he argues is a measure of the information inaccuracy, subject to Equations (3.10) and (3.11) as constraints. This can only be done approximately as the resulting equations are too difficult to solve exactly. This seems worth mentioning as it indicates the possibility of further extensions in our application of the concept of entropy.

Thus there is some evidence that when gravity models are introduced directly into certain economic models only the most rudimentary forms are used, even in association with more sophisticated techniques such as those developed by Theil. This supports the contention made in the introduction to this chapter that the geographic and economic approaches should be integrated.

3.2.3 Deficiencies of the gravity-model approach

All the gravity models which have been stated have been written down for individual commodity flows, and yet these flows must be interrelated. Part of the flow of commodity m from i to j could arise, for example, from the needs of other production sectors in j which use commodity m as an input. This difficulty can be overcome to some extent if the gravity-model variants used are those which incorporate independent estimates of X_i^m and Y_j^m. Suppose, for example, that regression analysis is used to estimate these quantities. The independent variables in the regression equations for X_i^m and Y_j^m could include measures of production and use in regions i and j of commodities other than m. The coefficients in such equations would play a similar role to input–output coefficients and effectively link the flow of models for different commodity groups. However, it is clear that such a procedure cannot reproduce the full richness of the input–output model in representing the economic interdependencies, and so this supports from the other side the contention that there is a need to integrate geographic and economic approaches. Firstly, however, in the next section entropy-maximising principles will be used to derive the family of gravity models discussed here. A general outline will then be given of input–output analysis, and finally it will be shown how the gravity-model and input–output approaches can be integrated using entropy-maximising methods.

3.3 Derivation of the family of gravity models using entropy-maximising methods

We have seen in Chapter 2 that the entropy-maximising principle offers a general tool. If a set of variables are to be estimated, such as the flows x_{ij}^m, and if the known constraints on x_{ij}^m can be expressed in equation form, then the entropy of a probability distribution associated with x_{ij}^m can be maximised and a maximum probability estimate of x_{ij}^m obtained. Before we use this general tool to integrate the gravity and input-output model approaches, it will be useful to show how the full range of gravity models presented in Section 3.2 can be derived, and this will further deepen our understanding of the gravity models themselves.

Model 8 of Table 3.1 [production-attraction-constrained Newtonian, case (iv)] has been derived in Chapter 2 for the person-trip case. We replace T_{ij} by x_{ij}^m, O_i by X_i^m, D_j by Y_j^m, c_{ij} by c_{ij}^m, and we assume that a total amount C^m is spent on transporting commodity m. Then we maximise

$$S^m = -\sum_i \sum_j \ln x_{ij}^m ! \tag{3.29}$$

subject to the constraint Equations (3.10) and (3.11), repeated here for convenience,

$$\sum_j x_{ij}^m = X_i^m \tag{3.10}$$

$$\sum_i x_{ij}^m = Y_j^m \ , \tag{3.11}$$

and a cost constraint [equivalent to Equation (2.8)],

$$\sum_i \sum_j x_{ij}^m c_{ij}^m = C^m \ . \tag{3.30}$$

This gives

$$x_{ij}^m = A_i^m B_j^m X_i^m Y_j^m \exp(-\mu^m c_{ij}^m) \tag{3.31}$$

where

$$A_i^m = \left[\sum_j B_j^m Y_j^m \exp(-\mu^m c_{ij}^m)\right]^{-1} \tag{3.32}$$

and

$$B_j^m = \left[\sum_i A_i^m X_i^m \exp(-\mu^m c_{ij}^m)\right]^{-1} \ . \tag{3.33}$$

This model is now equivalent to that given in Equations (3.10)-(3.18), but with the negative exponential function $\exp(-\mu^m c_{ij}^m)$ replacing the general function $f^m(c_{ij}^m)$. As in Chapter 2, the negative exponential function will be used henceforth, though the comments of that chapter relating to the possibilities of transforming the function should be borne in mind. Notice that we are using μ^m as the parameter in the negative exponential function, and not β^m which we would do if we presented a strict analogy of notation with Chapter 2. This allows us to continue to use α^m and β^m as powers for the 'mass' terms.

Cases (ii) and (iii) of the constrained Newtonian version can be derived in exactly the same way by repeating the case (iv) derivation, but firstly dropping Equation (3.11) as a constraint and then dropping Equation (3.10). However, the results are not exactly in the form of the equations listed in Table 3.1. Consider case (ii) as an example. If the procedure just outlined is followed, we obtain

$$x_{ij}^m = A_i^m X_i^m \exp(-\mu^m c_{ij}^m) \tag{3.34}$$

where

$$A_i^m = \left[\sum_j \exp(-\mu^m c_{ij}^m)\right]^{-1} . \tag{3.35}$$

There is no term in x_{*j}^m as there is in the corresponding Equations (3.12) and (3.13). However, such a term can be introduced using a device from an earlier paper (see Wilson, 1967a). Why do we want the x_{*j}^m term? The answer is that if we do not have one, then the product X_i^m is allocated among regions j in proportion to what can be called the transport accessibility of j to i, $\exp(-\mu^m c_{ij}^m)$. There may be other characteristics of region j (that is, other than accessibility ones) which make it beneficial for j to import commodity m rather than for some other region to do so. We can build in such a hypothesis as follows: let v_j^m be the benefit in j arising from the use of a unit of m, compared with the use of that unit in any other region. Choose units so that this can be offset directly against the transport cost c_{ij}^m. Thus Equations (3.34) and (3.35) can be rewritten as

$$x_{ij}^m = A_i^m X_i^m \exp(\mu^m v_j^m)\exp(-\mu^m c_{ij}^m) \tag{3.36}$$

and

$$A_i^m = \left[\sum_j \exp(\mu^m v_j^m)\exp(-\mu^m c_{ij}^m)\right]^{-1} . \tag{3.37}$$

Now, how would we actually estimate v_j^m? One possible argument is that such benefits as v_j^m are usually generated by scale economics in region j, and so a proxy variable would be the current scale of usage of commodity m in j, which is x_{*j}^m. The usual kind of assumption in such circumstances is that the benefits are proportional to the log of the scale, so

$$\mu^m v_j^m = \alpha^m \ln x_{*j}^m , \tag{3.38}$$

where α^m is an appropriate parameter. Equations (3.36) and (3.37) then transform into Equations (3.12) and (3.13) if we substitute for v_j^m from Equation (3.38). This completes our derivation for case (ii). Case (iii) can be treated similarly by introducing a benefit u_i^m which is proportionally related to $\ln x_{i*}^m$. So also can case (i), but this is perhaps different enough to be worth an explicit statement.

Case (i), listed as model 5 under the so-called constrained Newtonian models, is really a repeat of model 1. There are constrained variants, but it has already been explained that these reduce to slightly amended versions of cases (ii)–(iv). So the version of (i) we have to consider, in the new approach, has Equation (3.30) as a constraint together with Equation (3.4). This leads to

$$x_{ij}^m = K^m \exp(-\mu^m c_{ij}^m) , \tag{3.39}$$

where K^m is a normalising factor calculated, as it was for Equation (3.5), by using Equation (3.4). Terms in $(x_{i*}^m)^{\alpha^m}$ and $(x_{*j}^m)^{\beta^m}$ can be introduced as in cases (ii) and (iii), but with both of the benefits u_i^m and v_j^m being deemed to operate, to give

$$x_{ij}^m = K^m (x_{i*}^m)^{\alpha^m} (x_{*j}^m)^{\beta^m} \exp(-\mu^m c_{ij}^m). \tag{3.40}$$

Note that the versions of cases (i)–(iii) of the constrained Newtonian model which we have now derived are in fact models 13, 14, and 15 of Table 3.1, the constrained Newtonian models with 'masses' raised to a power where possible. Models 5, 6, and 7, of course, are simply special cases of these with α^m and/or β^m taken to be 1 as appropriate. Thus, we have now, in effect, shown how to derive the complete family of gravity models using entropy-maximising principles.

3.4 The input–output model

3.4.1 Principles of the single-region input–output model

3.4.1.1 *The economic-base model*

It is instructive to show how the input–output model for a single region grows out of, and can be contrasted with, the theory of the economic base (following Artle, 1961). It was pointed out in the introduction that the economic-base theory is still commonly used for local or regional economic forecasting. It represents the simplest possible statement of interdependence of economic sectors. Two sectors only of economic activity are defined and are usually measured in employment units. These sectors are known as basic employment and non-basic employment. Economic activity in the basic sector is supposed to be production for non-local markets (so it is the exporting sector), and the non-basic sector produces for local markets. Let X denote total employment, n non-basic employment, and y basic employment. Then

$$X - n = y , \tag{3.41}$$

which can be written

$$X\left(1 - \frac{n}{X}\right) = y . \tag{3.42}$$

Put

$$a = \frac{n}{X} , \tag{3.43}$$

so

$$X = (1-a)^{-1}y \ . \tag{3.44}$$

The usual assumption of the economic-base theory is that a is constant, so that, given an estimate of basic employment y, total employment can be estimated from Equation (3.44). $(1-a)^{-1}$ acts as a multiplier. Artle derived the economic-base theory in this form to be able to contrast it with input–output theory.

3.4.1.2 *The basic set of accounts*
Let us now revert to the situation defined at the beginning of the chapter where we have a set of commodity groups labelled m, n, p, etc. Let X^m be the total volume of production of commodity m (that is, we are using our previous notation, but for a single region so that regional subscripts are unnecessary). This product can be used as input to other sectors endogenous to the model or 'consumed' in an exogenous (final demand) sector. Suppose that Z^{mn} of X^m is used as an input to sector n, and y^m is consumed by the final demand sector. Then these quantities can be arranged in a system of accounts, as shown in Table 3.2. The commodity groups will be considered endogenous to the input–output model. Inputs are recorded in columns, outputs in rows. There is an exogenous sector which requires an additional row and column. The additional column is usually termed 'final demand' and includes exports as well as direct consumption by the household sector. The additional row consists of the outputs of factors of production, such as the household sector, measured in wages, and of imports. The superscript zero has been used for this sector. If financial units are being used, columns can be considered to contain payments and rows to contain receipts. One final column has been added to contain the X^m's, the total outputs, which can now be seen to be the row totals. Note that no declaration needs to be made about units for the purposes of this paper. It was stated in Section 3.1.3 that commodity quantities were measured in units appropriate to the commodity, and that comment still holds. Thus, any row of Table 3.2 contains quantities measured in the same units—the units of outputs of a particular commodity group. Since different units may be used for different rows, it is not possible to sum over columns. [If monetary units are used, then it is possible to sum over columns; receipts and payments can be equated for each commodity group, and the resulting relation implies something about prices. It is necessary to study such implications to develop a full economic model, but not the purposes of this chapter—see Dorfman, Samuelson, and Solow (1958) for a discussion of this point.] However,

rows can be summed, and it can easily be seen that the appropriate relationship is

$$\sum_n Z^{mn} + y^m = X^m \ . \tag{3.45}$$

Table 3.2. System of accounts for a single region.

		Inputs to commodity (payments)						Exogenous sector (final demand)	Total output
		1	2	3	. . .	n	. . .		
Outputs from commodity (receipts)	1	Z^{11}	Z^{12}	Z^{13}	. . .	Z^{1n}	. . .	y^1	X^1
	2	Z^{21}	Z^{22}	Z^{23}	. . .	Z^{2n}	. . .	y^2	X^2
	3	Z^{31}	Z^{32}	Z^{33}	. . .	Z^{3n}	. . .	y^3	X^3
	.				.			.	.
	.				.			.	.
	.				.			.	.
	m	Z^{m1}	Z^{m2}	Z^{m3}	. . .	Z^{mn}		y^m	X^m
	.				.			.	.
	.				.			.	.
	.				.			.	.
Exogenous sector (factors of production, imports, etc.)		Z^{01}	Z^{02}	Z^{03}	. . .	Z^{0n}			

3.4.1.3 *The single-region input–output model*

Equation (3.45) is an identity, and we must make a hypothesis to turn it into an equation which makes a positive assertion and which can form the basis of a model. This can be done as follows: Z^{mn} in Equation (3.45) can be replaced by $(Z^{mn}/X^n)X^n$. However, the hypothesis can now be made that

$$a_{mn} = \frac{Z^{mn}}{X^n} \tag{3.46}$$

is a constant coefficient—that is, that the quantity of commodity m needed to produce a unit of commodity n is constant. Equation (3.45) can then be written as

$$\sum_n a_{mn} X^n + y^m = X^m \tag{3.47}$$

which can be solved for X^m and written in matrix notation without indices as

$$X = (I-a)^{-1}y \, . \tag{3.48}$$

a is known as the matrix of technical input–output coefficients, where I is the unit matrix.

Equation (3.48) can now be contrasted with Equation (3.44). They appear to be the same, but X, a, and y are scalar quantities in Equation (3.44), but vectors and matrices in Equation (3.48). In deriving Equation (3.48) from the set of accounts in Table 3.2, it can be seen that the input–output model is based on a more detailed description of economic structure than is implicit in the economic-base model. So, although Equation (3.48) is based on a rather strong assumption (that a_{mn} is constant), it must be expected to give a much more realistic account of the economic structure of a city or region than the economic-base model.

3.4.1.4 *Difficulties associated with the single-region model*
Artle (1961) points out that exports, which are so significant in the base model, form only a part of the exogenous sector of the input–output model, and not the largest part of that. The largest part of the exogenous sector is the household sector and Artle (see, especially, Artle, 1965) has expanded the endogenous component of the input–output model by including the household sector endogenously. However, there could still be difficulties in forecasting the exogenous sector when the model is used for forecasting purposes.

The second major difficulty of the input–output model, as derived so far, is the main topic of this chapter: how to extend the model to represent a multiregional economy. In the form presented in Section 3.4.1.3, the input–output model exposes the weakness of the economic-base approach, but it cannot yet be argued that it is an adequate replacement. It is rather difficult to use as a forecasting model because of the extensive exogenous sector, and, because a city or region is part of a multiregional economy, complex import–export substitutions can take place between different regions which can affect the final demand for exports in any one region in a complex way. So the extension of the model to cover the multiregional case is vital. The framework to be built on here is that developed by Leontief and Strout (1963).

3.4.2 The Leontief–Strout multiregional framework

The framework developed by Leontief and Strout rests on one rather elegant assumption: that the ultimate destination of goods is irrelevant to producers, and that the origin of goods is irrelevant to consumers. This implies that all goods produced in region i can be considered to pass into a supply pool, and all goods used in i to be extracted from a demand pool. The relevant flows in such a system are shown in Figure 3.1. The notation used in Figure 3.1, and henceforth, is that developed earlier, and is related to that used for the single input–output model with regional indices added. Note that, for a single region i, X_i^m would be equal to Y_i^m, but in a multiregional system this need not be the case. x_{ij}^m is now the quantity of commodity m produced in region i and shipped (notionally) to a demand pool in region j.

Each region individually must now satisfy an input–output equation of the form of Equation (3.47). It can easily be cbecked, with the help of Figure 3.1, that the appropriate set of equations can be written as

$$x_{*i}^m = \sum_n a_{mn}^i x_{i*}^n + y_i^m \tag{3.49}$$

where (recapping) x_{*i}^m is the amount of commodity m used in region i; y_i^m is the consumption by the final demand sector in region i; $\sum_n a_{mn}^i x_{i*}^n$ is the quantity of commodity m used by other sectors (and in this expression a_{mn}^i is the matrix of technical input–output coefficients for region i). This is the form in which the equation is given by Leontief and Strout. They take x_{ij}^m to be the basic variables to be solved for in the model and adopt weaker assumptions than any we considered above: that not only are x_{i*}^m and x_{*j}^m not independently estimated as X_i^m and Y_j^m, but no independent estimate is available for X^m either. They do, however, make a gravity-model type of assumption as already discussed in Section 3.2.2.3 and presented in Equation (3.26), but they assume that Equation (3.26) holds only for $i \neq j$. Then, with this restriction, Equations (3.20) and (3.49) give just as many equations as unknowns (including the diagonal

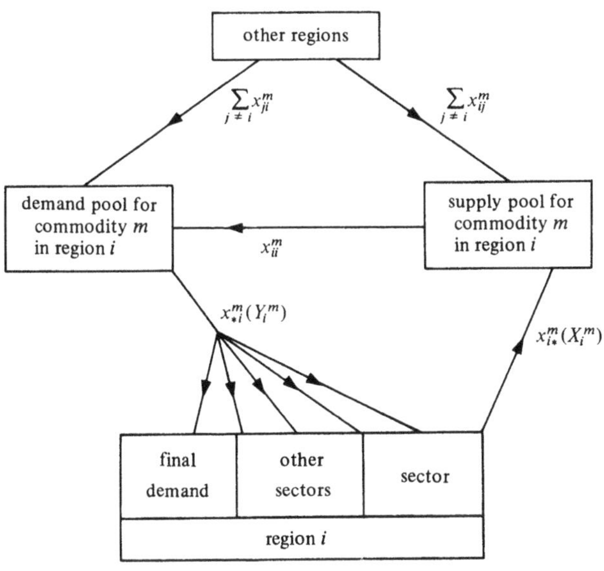

Figure 3.1. Flows of commodity m in the Leontief–Strout framework.

terms x_{ii}^m, of course) and Leontief and Strout discuss various ways of solving these equations. The equation system given by Equations (3.26) and (3.49) is nonlinear in x_{ij}^m. The equations are solved approximately by assuming that all the flows are known for some base year, and then the equations can be solved (to the first order) for the incremental flows up to a subsequent year. That is, given a set of increments in final demand, the increments in the flows can be determined to the first order. Such procedures are usually only useful for very short-term forecasting, as they represent a sophisticated method of trend projection. However, the details of the solution can be studied in the original paper by Leontief and Strout. It is clear from our earlier assumptions that more general gravity models could be used to replace Equation (3.26). We shall now try to go further, however, and integrate the gravity and input–output model approaches by using entropy-maximising principles.

3.5 Integrating the gravity and input–output model approaches

3.5.1 Introduction

We shall consider separately each of the situations we have termed cases (i)–(iv) and we shall adopt the simplifying assumption of Section 3.2.1.2 about which constraints we shall consider in each case: only those which include an X_i^m or Y_j^m term on the right-hand side. All other possibilities are variants of the four which will now be considered and appropriate models could be generated if desired.

3.5.2 Case (i): the unconstrained gravity input–output model

In the gravity-model approach to case (i), we assumed that an independent estimate of X^m did exist, though there was no such estimate for X_i^m and Y_j^m. For this particular development of the integrated model, we shall assume that there is no such estimate of X^m, and this brings us into line with the assumptions of Leontief and Strout. Thus the case (i) model to be developed here represents the modification to the Leontief–Strout model brought about by integrating the gravity and input–output models using entropy-maximising principles, but otherwise no new assumptions are made. The only constraints, then, are Equation (3.30), which is restated here for convenience:

$$\sum_i \sum_j x_{ij}^m c_{ij}^m = C^m \ , \tag{3.30}$$

and we now treat the Leontief–Strout equation (3.49) as a constraint on x_{ij}^m, and so we rewrite it in terms of the x_{ij}^m themselves as

$$\sum_j x_{ji}^m = \sum_n a_{mn}^i \sum_j x_{ij}^n + y_i^m \ . \tag{3.50}$$

We now have to maximise the entropy of the probability distribution associated with x_{ij}^m (with m varying now as well as i and j). It can easily be seen that there is no need to normalise x_{ij}^m to obtain a probability distribution, as the same estimate is obtained if we maximise our entropy S defined by[(7)]

$$S = -\sum_i \sum_j \sum_m x_{ij}^m \ln x_{ij}^m \tag{3.51}$$

subject to Equations (3.30) and (3.50). To do this we form the Lagrangian $\mathcal{L}$, where

$$\mathcal{L} = S + \sum_i \sum_m \gamma_i^m \left(y_i^m + \sum_n a_{mn}^i \sum_j x_{ij}^n - \sum_j x_{ji}^m \right) + \sum_m \mu^m \left(C^m - \sum_i \sum_j x_{ij}^m c_{ij}^m \right) , \tag{3.52}$$

where γ_i^m is the set of Lagrangian multipliers associated with Equation (3.50) and μ^m the set associated with Equation (3.30). We now obtain an estimate of x_{ij}^m by solving

$$\frac{\partial \mathcal{L}}{\partial x_{ij}^m} = 0 \tag{3.53}$$

together with the constraint Equations (3.30) and (3.50). Equation (3.53) gives

$$x_{ij}^m = \exp\left(\sum_n \gamma_i^n a_{nm}^i - \gamma_j^m - \mu^m c_{ij}^m \right) \tag{3.54}$$

(where a 1 has been absorbed into the multiplier γ_j^m, without loss of generality). μ^m is obtained by substituting x_{ij}^m into Equation (3.30), or, if C^m is not known explicitly, by calibrating in the usual way. γ_j^m is found by substituting x_{ij}^m from Equation (3.54) into Equation (3.50). This gives

$$\exp(-\gamma_i^m) \sum_j \exp\left(\sum_n \gamma_j^n a_{nm}^j - \mu^m c_{ji}^m \right) - \sum_n a_{mn}^i \exp\left(\sum_p \gamma_i^p a_{pn}^i \right) \sum_j \exp(-\gamma_j^n - \mu^m c_{ij}^n) - y_i^m = 0 . \tag{3.55}$$

Define

$$\delta_i^m = \exp\left(\sum_n \gamma_i^n a_{nm}^i \right) \tag{3.56}$$

and

$$\epsilon_j^m = \exp(-\gamma_j^m) , \tag{3.57}$$

so that

$$\delta_i^m = \prod_n (\epsilon_i^n)^{-a_{mn}^i} \tag{3.58}$$

(7) It is convenient in this case to use this form of S. The use of $-\ln x_{ij}^m!$ appears to cause conceptual difficulties if x_{ij}^m is noninteger. This procedure avoids any such difficulties. It can easily be checked that S, above, gives the same answer as using x_{ij}^m / X^m, the normalised form, which is strictly correct if S is to be defined as the entropy of a probability distribution.

Then Equation (3.55) can be written more tidily as

$$\epsilon_i^m \sum_j \delta_j^m \exp(-\mu^m c_{ji}^m) - \sum_n a_{mn}^i \delta_i^n \sum_j \epsilon_j^n \exp(-\mu^n c_{ij}^n) - y_i^m = 0 \,, \tag{3.59}$$

which can be rearranged to give for ϵ_i^m:

$$\epsilon_i^m = \frac{y_i^m + \sum_n a_{mn}^i \delta_i^n \sum_j \epsilon_j^n \exp(-\mu^n c_{ij}^n)}{\sum_j \delta_j^m \exp(-\mu^m c_{ij}^m)} \,. \tag{3.60}$$

The equation cannot, of course, be solved explicitly for ϵ_i^m, but Equation (3.60) suggests a suitable iterative procedure: guess ϵ_i^m, calculate δ_i^m from Equation (3.58), recalculate ϵ_i^m from Equation (3.60), and continue this cycle until the process converges. Then, using Equations (3.56) and (3.57), the Equation (3.54) for x_{ij}^m can be written as

$$x_{ij}^m = \delta_i^m \epsilon_j^m \exp(-\mu^m c_{ij}^m). \tag{3.61}$$

Thus the entropy-maximising model for what might be called the Leontief–Strout version of our case (i) is given by Equations (3.61), (3.60), and (3.58).

A number of observations can be made about Equation (3.61). The product $\delta_i^m \epsilon_j^m$ ensures that the flows satisfy the Leontief–Strout equation. Note from Equation (3.60) that the dimension of this product is that unit which is used to measure quantities of the commodity m—this can be seen from the role played by the final demand term y_i^m in Equation (3.60), and this must, of course, be supplied exogenously. Note also that the gravity-model behaviour in Equation (3.61) is the simplest possible: x_{ij}^m is proportional to $\exp(-\mu^m c_{ij}^m)$, but not to any other particular characteristics of region i or j. As we have assumed that independent estimates of X_i^m and Y_j^m do not exist, we cannot reasonably introduce terms in these variables. However, terms $(x_{i*}^m)^{\alpha^m}$ and $(x_{*j}^m)^{\beta^m}$ could be introduced using the same device that led to Equation (3.40) in Section 3.3. This is done simply by replacing $\exp(-\mu^m c_{ij}^m)$ by the product $(x_{i*}^m)^{\alpha^m}(x_{*j}^m)^{\beta^m}\exp(-\mu^m c_{ij}^m)$ and $\exp(-\mu^m c_{ji}^m)$ by the same expression with j and i reversed, wherever either expression appears in Equations (3.58), (3.60), and (3.61).

3.5.3 Case (ii): the production-constrained gravity input–output model

Now consider case (ii): we assume that we have an independent estimate of X_i^m, but still not of Y_j^m. The constraints are then Equations (3.10), (3.30), and (3.50), but, in Equation (3.50), $\sum_j x_{ij}^m$ should be replaced by X_i^m. Thus the appropriate Lagrangian for this case is

$$\mathcal{L} = -\sum_i \sum_j \sum_m x_{ij}^m \ln x_{ij}^m + \sum_i \sum_m \gamma_i^m \left(y_i^m + \sum_n a_{mn}^i X_i^n - \sum_j x_{ji}^m \right) + \sum_i \sum_m \lambda_i^{(1)m} \left(X_i^m - \sum_j x_{ij}^m \right) + \sum_m \mu^m \left(C^m - \sum_i \sum_j x_{ij}^m c_{ij}^m \right) . \tag{3.62}$$

This leads to the following estimate of x_{ij}^m:

$$x_{ij}^m = \exp(-\lambda_i^{(1)m} - \gamma_j^m - \mu^m c_{ij}^m) \, . \tag{3.63}$$

We can solve for μ^m using Equation (3.30) or by calibration in the usual way, for $\lambda_i^{(1)m}$ from Equation (3.10), and for γ_j^m from Equation (3.50), but with X_i^m replacing $\sum_j x_{ij}^m$. Put

$$\exp(-\lambda_i^{(1)m}) = A_i^m X_i^m \, , \tag{3.64}$$

and henceforth use A_i^m instead of $\lambda_i^{(1)m}$. Then Equation (3.10) gives

$$A_i^m = \left[\sum_j \exp(-\gamma_j^m - \mu^m c_{ij}^m)\right]^{-1} \, . \tag{3.65}$$

We can again define ϵ_i^m using Equation (3.57), and then the amended Equation (3.50) gives

$$\epsilon_i^m = \frac{y_i^m + \sum_n a_{mn}^i X_i^n}{\sum_j A_j^m X_j^m \exp(-\mu^m c_{ji}^m)} \tag{3.66}$$

and Equation (3.63) can be written

$$x_{ij}^m = A_i^m X_i^m \epsilon_j^m \exp(-\mu^m c_{ij}^m) \, , \tag{3.67}$$

and so Equations (3.67), (3.60), and (3.65) give the entropy-maximising model for case (ii). Note that these equations could be modified to introduce a term $(x_{*j}^m)^{\beta^m}$ as a factor by the device of Section 3.3 if required.

3.5.4 Case (iii): the attraction-constrained gravity input-output model

It can now easily be seen that the case (iii) result can be obtained similarly, and the result is

$$x_{ij}^m = \delta_i^m B_j^m Y_j^m \exp(-\mu^m c_{ij}^m) \, , \tag{3.68}$$

where δ_i^m is obtained by solving Equation (3.50) written in the form

$$\sum_n a_{mn}^i \delta_i^n = Y_i^m - y_i^m \, , \tag{3.69}$$

which is a set of linear equations in δ_i^n for each region i. B_j^m is obtained from Equation (3.11) as

$$B_j^m = \left[\sum_i \delta_i^m \exp(-\mu^m c_{ij}^m)\right]^{-1} \, . \tag{3.70}$$

In this case a factor $(x_{i*}^m)^{\alpha^m}$ can be inserted if desired using the device of Section 3.3.

3.5.5 Case (iv): the production-attraction-constrained gravity input-output model

Finally, we have to consider case (iv). Equations (3.10) and (3.11) are now both constraints, along with Equations (3.30) and (3.50) amended, so that X_i^m replaces $\sum_j x_{ij}^m$ and Y_j^m replaces $\sum_i x_{ij}^m$ to give

$$Y_i^m = \sum_n a_{mn}^i X_i^n + y_i^m \quad . \tag{3.71}$$

This reveals that we have a very simple solution in this case. x_{ij}^m does not now appear in Equation (3.71), and so our assumptions have decoupled the input–output part of the model from the gravity-model part. This means that the procedures which are used to estimate X_i^m and Y_j^m should be such that Equation (3.71) is satisfied, and the flows x_{ij}^m are then estimated using the case (iv) gravity model given by Equations (3.31)–(3.33).

3.5.6 A hybrid model for different commodity types

It is clear that different types of good will fall into different categories. For example, a primary commodity, such as coal, is likely to be production-constrained; a commodity which is mainly an intermediate good for primary sectors would be attraction-constrained; a primary commodity which is an input for other primary sectors would be production-attraction-constrained; and there are a wide variety of goods which are neither production- nor attraction-constrained. The models developed so far, in Sections 3.5.2–3.5.5, could each only be applied to a subset of commodities. Thus we need a hybrid model. We can now show that entropy-maximising principles can again be used to generate such a model. We shall retain the simplifying assumption that constraints will only be applied if there is an X_i^m or Y_j^m term on the right-hand side (rather than x_{i*}^m or x_{*j}^m).

Let M_x be a set of commodities to which the assumptions of case (x) apply. Denote by $m \in M_i$ the fact that commodity m is in set i, and let $\sum_{m \in M_i}$ denote summation over all $m \in M_i$. Then the constraints on x_{ij}^m can be seen to be as follows:

$$\sum_j x_{ij}^m \quad = X_i^m \, , \qquad m \in M_2 \, , M_4 \tag{3.72}$$

$$\sum_i x_{ij}^m \quad = Y_j^m \, , \qquad m \in M_3 \, , M_4 \tag{3.73}$$

$$\sum_i \sum_j x_{ij}^m c_{ij}^m = C^m \tag{3.74}$$

$$\sum_j x_{ji}^m - \sum_n a_{mn}^i \sum_j x_{ij}^n - y_i^m = 0 \, , \qquad m \in M_1 \tag{3.75}$$

$$\sum_j x_{ji}^m - \sum_n a_{mn}^i X_i^n - y_i^m \quad = 0 \, , \qquad m \in M_2 \tag{3.76}$$

$$Y_i^m - \sum_n a_{mn}^i \sum_j x_{ij}^n - y_i^m \quad = 0 \, , \qquad m \in M_3 \tag{3.77}$$

and

$$Y_i^m - \sum_n a_{mn}^i X_i^n - y_i^m = 0 , \qquad m \in M_4 . \tag{3.78}$$

The corresponding Lagrangian is

$$\begin{aligned}\mathcal{L} = & -\sum_i \sum_j \sum_m x_{ij}^m \ln x_{ij}^m + \sum_i \sum_{m \in M_2, M_4} \lambda_i^{(1)m} \left(X_i^m - \sum_j x_{ij}^m \right) \\ & + \sum_j \sum_{m \in M_3, M_4} \lambda_j^{(2)m} \left(Y_j^m - \sum_i x_{ij}^m \right) + \sum_m \mu^m \left(C^m - \sum_i \sum_j x_{ij}^m c_{ij}^m \right) \\ & + \sum_i \sum_{m \in M_1} \gamma_i^{(1)m} \left(y_i^m + \sum_n a_{mn}^i \sum_j x_{ij}^n - \sum_j x_{ji}^m \right) \\ & + \sum_i \sum_{m \in M_2} \gamma_i^{(3)m} \left(y_i^m + \sum_n a_{mn}^i X_i^n - \sum_j x_{ji}^m \right) \\ & + \sum_i \sum_{m \in M_3} \gamma_i^{(3)m} \left(y_i^m + \sum_n a_{mn}^i \sum_j x_{ij}^n - Y_i^m \right) .\end{aligned} \tag{3.79}$$

Notice that the main effect of our new hybrid assumptions are that $\lambda_i^{(1)m}$ and $\lambda_j^{(2)m}$ are each defined over a restricted range of m, and what was γ_i^m has now three components $\gamma_i^{(1)m}$, $\gamma_j^{(2)m}$, and $\gamma_j^{(3)m}$, each defined for restricted ranges of m. Note also that the constraint Equation (3.78) is not represented in the Lagrangian as it does not contain x_{ij}^m. The resulting entropy-maximising model is obtained in the usual way by solving

$$\frac{\partial \mathcal{L}}{\partial x_{ij}^m} = 0 \tag{3.80}$$

together with the constraint Equations (3.72)–(3.78). From Equation (3.80), for the various sets M_i, we obtain

$$x_{ij}^m = \exp\left(\sum_n a_{mn}^i \gamma_i^{(1)n} - \gamma_j^{(1)m} - \mu^m c_{ij}^m \right) , \qquad m \in M_1 \tag{3.81}$$

$$x_{ij}^m = \exp\left(-\lambda_i^{(1)m} - \gamma_j^{(2)m} - \mu^m c_{ij}^m \right) , \qquad m \in M_2 \tag{3.82}$$

$$x_{ij}^m = \exp\left(\sum_n a_{mn}^i \gamma_i^{(3)n} - \lambda_j^{(2)m} - \mu^m c_{ij}^m \right) , \qquad m \in M_3 \tag{3.83}$$

$$x_{ij}^m = \exp\left(-\lambda_i^{(1)m} - \lambda_j^{(2)m} - \mu^m c_{ij}^m \right) . \qquad m \in M_4 \tag{3.84}$$

Put

$$\exp(-\lambda_i^{(1)m}) = A_i^m X_i^m , \qquad m \in M_2 , M_4 \tag{3.85}$$

$$\exp(-\lambda_j^{(2)m}) = B_j^m Y_j^m , \qquad m \in M_3 , M_4 \tag{3.86}$$

$$\delta_i^{(1)m} = \exp\left(\sum_n a_{mn}^i \gamma_i^{(1)n} \right) , \qquad m \in M_1 \tag{3.87}$$

$$\delta_i^{(3)m} = \exp\left(\sum_n a_{mn}^i \gamma_i^{(3)n} \right) , \qquad m \in M_3 \tag{3.88}$$

$$\epsilon_j^{(1)m} = \exp(-\gamma_j^{(1)m}), \qquad m \in M_1 \tag{3.89}$$

$$\epsilon_j^{(2)m} = \exp(-\gamma_j^{(2)m}), \qquad m \in M_2 \tag{3.90}$$

so that

$$\delta_i^{(1)m} = \prod_n (\epsilon_i^{(1)n})^{-a^i_{mn}} \ . \tag{3.91}$$

We can now solve for A_i^m, B_j^m, μ^m, $\epsilon_i^{(1)m}$, $\epsilon_j^{(2)m}$ and $\delta_i^{(3)m}$ by substituting in Equations (3.72), (3.73), (3.74), (3.75), (3.76), and (3.77), respectively. X_i^m and Y_i^m must be related by Equation (3.78) for $m \in M_4$. Firstly, though, let us specify x_{ij}^m in terms of the variables just defined:

$$x_{ij}^m = \delta_i^{(1)m}\epsilon_j^{(1)m} \exp(-\mu^m c_{ij}^m), \quad m \in M_1 \tag{3.92}$$

$$x_{ij}^m = A_i^m X_i^m \epsilon_j^{(2)m} \exp(-\mu^m c_{ij}^m), \quad m \in M_2 \tag{3.93}$$

$$x_{ij}^m = \delta_i^{(3)m} B_j^m Y_j^m \exp(-\mu^m c_{ij}^m), \quad m \in M_3 \tag{3.94}$$

$$x_{ij}^m = A_i^m X_i^m B_j^m Y_j^m \exp(-\mu^m c_{ij}^m), \quad m \in M_4 \ . \tag{3.95}$$

Then we can use Equations (3.72)-(3.77), in conjunction with Equations (3.92)-(3.95), to give

$$A_i^m = \left[\sum_j \epsilon_j^{(2)m} \exp(-\mu^m c_{ij}^m)\right]^{-1}, \qquad m \in M_2 \tag{3.96}$$

$$A_i^m = \left[\sum_j B_j^m Y_j^m \exp(-\mu^m c_{ij}^m)\right]^{-1}, \qquad m \in M_4 \tag{3.97}$$

$$B_j^m = \left[\sum_i \delta_i^{(3)m} \exp(-\mu^m c_{ij}^m)\right]^{-1}, \qquad m \in M_3 \tag{3.98}$$

$$B_j^m = \left[\sum_i A_i^m X_i^m \exp(-\mu^m c_{ij}^m)\right]^{-1}, \qquad m \in M_4 \tag{3.99}$$

$$\begin{aligned}\epsilon_i^{(1)m} = y_i^m + \Bigg[& \sum_{n \in M_1} a^i_{mn} \delta_i^{(1)m} \sum_j \epsilon_j^{(1)n} \exp(-\mu^n c_{ij}^n) \\ & + \sum_{n \in M_2} a^i_{mn} A_i^n X_i^n \sum_j \epsilon_j^{(2)n} \exp(-\mu^n c_{ij}^n) \\ & + \sum_{n \in M_3} a^i_{mn} \delta_i^{(3)n} \sum_j B_j^n Y_j^n \exp(-\mu^n c_{ij}^n) \\ & + \sum_{n \in M_4} a^i_{mn} A_i^n X_i^n \sum_j B_j^n Y_j^n \exp(-\mu^n c_{ij}^n) \Bigg] \\ & \times \left[\sum_j \delta_j^{(4)m} \exp(-\mu^m c_{ji}^m)\right]^{-1}\end{aligned} \tag{3.100}$$

$$\epsilon_i^{(2)m} = \frac{y_i^m + \sum_n a^i_{mn} X_i^n}{\sum_j A_j^m X_j^m \exp(-\mu^m c_{ji}^m)} \tag{3.101}$$

$$\sum_{n \in M_3} a^i_{mn} \delta_i^{(3)m} \sum_j B^n_j Y^n_j \exp(-\mu^n c^n_{ij})$$
$$= Y^m_i - y^m_i - \sum_{n \in M_1} a^i_{mn} \delta_i^{(1)n} \sum_j \epsilon_j^{(1)n} \exp(-\mu^n c^n_{ij})$$
$$- \sum_{n \in M_2} a^i_{mn} A^n_i X^n_i \sum_j \epsilon_j^{(2)n} \exp(-\mu^n c^n_{ij})$$
$$- \sum_{n \in M_4} a^i_{mn} A^n_i X^n_i \sum_j B^n_j Y^n_j \exp(-\mu^n c^n_{ij}) \ . \tag{3.102}$$

Thus the hybrid model is given by Equations (3.92)–(3.102). This equation system can be solved as follows: guess initial values of A^m_i, $\delta_i^{(1)m}$, and $\delta_i^{(3)m}$; calculate B^m_j from Equations (3.98) and (3.99); calculate $\epsilon_i^{(2)m}$ from Equation (3.101); calculate A^m_i from Equations (3.96) and (3.97); calculate $\epsilon_i^{(1)m}$ from Equation (3.100); calculate $\delta_i^{(1)m}$ from Equation (3.91), and recalculate $\epsilon_i^{(1)m}$; calculate $\delta_i^{(3)m}$ by solving the linear equation system, Equation (3.102); repeat until the process converges. Although this sounds complex, experience with, for example, transport models, suggests that this sort of iterative procedure converges quite rapidly. Note that terms $(x^m_{i*})^{\alpha^m}$ [or $(x^m_{*j})^{\beta^m}$] could be introduced into the equations in cases where there is no X^m_i (or Y^m_j) term in the usual way.

3.6 Summary

We have studied a system of interregional commodity flows x^m_{ij}. It has been shown that it is especially important to decide whether the quantities x^m_{i*} and x^m_{*j} can be estimated independently (as X^m_i and Y^m_j) of the procedure which estimates x^m_{ij}. There are a number of possible assumptions, and different models are developed, within a number of types of approach, for each set of assumptions. Thus one of the most important conclusions to be drawn from this chapter is that the hypotheses which are being represented in models should be examined as deeply and explicitly as possible.

It has been shown in Section 2 that, even with the conventional approach to the gravity model, a wide variety of models can be developed. The economic difficulties associated with gravity models have been pointed out, and the input–output framework has been outlined to show another approach to economic structure. It is shown in Section 3.3 that the whole family of gravity models can be derived by applying entropy-maximising principles to the various sets of possible assumptions. We can then see that the interregional input–output equations can be treated as constraints, and a number of integrated gravity and input–output models are demonstrated. Finally, we developed a general hybrid model, recognising that there are particular commodities whose movement is in accord with any one set of assumptions considered. *Any one of the other models derived in this paper could be shown to be special cases of this general model.*

The principle conclusion of this chapter is perhaps that there is no simple model which will adequately represent systems as complex as the ones we have discussed. Many existing models have been derived within more general families of models. The methods outlined here should enable the researcher to focus on any particular set of assumptions (taking account of his data limitations) he wants to test and so construct the appropriate model.

Finally, it should be mentioned that further extensions could be made to the models presented to take account of the fact that transport systems consist of several modes. The methods of two earlier papers (Wilson, 1967a, 1969a) could be used to achieve such extensions.

4

Location models

4.1 Gravity models as location models

In Chapters 2 and 3 we have constructed various kinds of transport flow models for both person trips and flows of goods. The discussion of families of gravity models in Chapter 3 implicitly introduces the concept of the gravity model as a location model, and we now take this notion forward explicitly. Suppose T_{ij} is a flow and we are using entropy-maximising principles to build a model of T_{ij}. If T_{ij} represents the journey to work for a homogeneous population, for example, then the gravity model given by Equations (2.21)–(2.23) of Chapter 2 would represent such a flow. It is usually assumed (and was assumed in Chapter 2) that independent estimates of the row and column sums of T_{ij}, T_{i*}, and T_{*j} exist. We called them O_i and D_j and used this information in building the model. We saw in Chapter 3, however, that there are cases where independent estimates of either T_{i*} or T_{*j} (or possibly, but more rarely, both) are unavailable. Suppose, for example, that T_{*j} is unavailable: then the corresponding gravity model (we called it the production-constrained model in Chapter 3) will itself provide an estimate of T_{*j}. Thus, should T_{ij} represent the interaction between residents of i and shops in j (say as a flow of cash expenditure on retail goods), then the model will provide an estimate of T_{*j}, the total amount of shopping activity in j. Such information is obviously useful in the planning of shopping centres.

Thus, in the cases where T_{ij} represents an interaction between two activities and at least one end of the interaction is unconstrained, the corresponding model provides an estimate of either T_{i*} or T_{*j}, the distribution of the activity by location at the unconstrained end. Some reservations must be attached to this general statement, however: if one end of the flow is unconstrained in the model simply because of lack of knowledge, then it may not be correct to use the flow model as a location model, as the lack of knowledge may generate a model which is a poor predictor of flows. We shall explore this point in more depth in relation to particular examples, and it is also instructive to relate such an exploration to existing practice in urban and regional modelling.

Three examples will be discussed: recreation, shopping, and residential location.

4.2 Recreation

It is useful to examine certain kinds of recreational systems as the first example for two reasons: at least hypothetically, it is possible to define a particularly simple type of system for analysis, and this facilitates a much deeper analysis of the underlying mathematics later in Chapter 7 and Appendix 2. We can also use and develop the formalism of Jaynes which was introduced in Section 1.2.2 of Chapter 1.

Consider a town with n residents who, on a particular weekend, each want to make an excursion to some recreational facility. Suppose the recreational facilities they can use are exclusive to the residents of this town, and that the expenditure incurred by an individual using the ith facility (say travel plus admission costs) is ϵ_i. This expenditure is assumed to be independent of the location of the residence of the individual in the town. We can enumerate all the ϵ_i's and we know the average expenditure $\bar{\epsilon}$. What is our best estimate of the number of residents n_i who use the ith facility?

Define

$$p_i = \frac{n_i}{n} , \tag{4.1}$$

and we can interpret p_i as the probability that an individual will use the ith facility. We know that

$$\sum_i p_i = 1 \tag{4.2}$$

and also that

$$\sum_i p_i \epsilon_i = \bar{\epsilon} . \tag{4.3}$$

Our problem is now exactly that of Section 1.1.2, and Equations (4.1) and (4.2) are equivalent to Equations (1.11) and (1.10), respectively, if we define $f(x_i)$ to be ϵ_i. We can combine Equations (1.13) and (1.14) by eliminating λ to obtain as a best estimate for p_i

$$p_i = \frac{\exp(-\mu\epsilon_i)}{\sum_i \exp(-\mu\epsilon_i)} , \tag{4.4}$$

and so

$$n_i = \frac{n\exp(-\mu\epsilon_i)}{\sum_i \exp(-\mu\epsilon_i)} . \tag{4.5}$$

We shall see in the next section that this is a special case of the production-constrained gravity model, used as a location model, and we shall return to it in Chapter 7 and Appendix 2.

The results given in Equations (4.4) and (4.5) are, of course, the same in form as the modal-split equations in Chapter 2. It should also be noted that the same form of probability distribution can be obtained from the study of 'logit regressions', as in Section 3.7 of Theil (1967).

4.3 Shopping

To discuss shopping models, it is useful to change notation again to emphasise that we are dealing with another specific example. The following are defined:

S_{ij} is the flow of expenditure from residents of zone i to shops in zone j;
e_i is the average expenditure of residents of zone i on shopping goods;
P_i is the population of zone i;
W_j is the weight to be associated with zone j as a proxy for shopping attractiveness;
c_{ij} is the 'cost' of travel from i to j.

The system is usually considered to be production-constrained, which means that

$$\sum_j S_{ij} = e_i P_i \ , \tag{4.6}$$

and S_{*j} is obtained from the model estimate of S_{ij}. Indeed, the principal application of such a model is to estimate S_{*j}, the turnover of particular shopping centres.

The production-constrained model which is usually used (cf. Lakshmanan and Hansen, 1965) is

$$S_{ij} = A_i(e_i P_i)W_j^{\alpha} \exp(-\beta c_{ij}) \ , \tag{4.7}$$

where

$$A_i = \left[\sum_j W_j^{\alpha} \exp(-\beta c_{ij})\right]^{-1} \ . \tag{4.8}$$

The reader can easily verify that this model can be derived using entropy-maximising principles, and is indeed a version of the production-constrained model described in Section 3.3 of Chapter 3 with appropriate changes in notation. The device introduced in Section 3.3 to bring in the W_j^{α} term [cf. Equation (3.38)] was in fact first introduced by the author (Wilson, 1967a) in connection with shopping models.

We now have to return to the point raised in Section 4.1 about whether the dropping of a constraint, in this case the attraction constraint, is merely lack of knowledge, or is a reasonable behavioral assumption. What is the implicit behavioral assumption in this case? It is that, given a distribution of purchasing power, shops will develop at places which are most accessible to this purchasing power [through the $\exp(-\beta c_{ij})$ term], due account being taken of scale economies (through the W_j^{α} term), and that no other 'constraint' on S_{*j} exists. This seems a reasonable assumption in this case, and is supported by the fact that models of this form fit existing situations reasonably well.

Many other developments should be mentioned in relation to shopping models (cf. Cordey Hayes, 1968). As the system being modelled is a production-constrained system, the approach of the intervening-opportunities model, essentially a production-constrained model, may be appropriate. This possibility and some related developments are discussed in Appendix 3.

We have also seen in Chapter 2 that any transport model of person flows should be disaggregated by mode and person type (car owner or not), and that the pattern of flows will then be determined within the model as a function of the availability of modes and their characteristics, and by car-ownership structure. In Chapter 2, these principles were applied within a production–attraction-constrained situation, the journey to work. It is clear that the same principles could be applied to the shopping model outlined above, and this is left as an exercise for the reader. This disaggregation would be particularly important, for example, in the investigation of out-of-town shopping centres where access would be primarily by car, in contrast to central city shopping centres where public transport access is still feasible. It should be noted, however, that shopping interaction in a production–attraction-constrained transport model is measured in terms of person trips, while in the production-constrained case discussed above, interaction is measured in money units of expenditure. The relationship between the two models is probably not simple, as longer person trips probably involve greater expenditure relatively.

This leads to one final point: presentation in this chapter has assumed that there is only one type of shopping good (or alternatively that all shopping goods are homogeneous with respect to the shopping behaviour they generate). It is clearly better to disaggregate and to run the model separately for a number of classes of good, making distinctions, for example, between convenience goods, such as groceries, and durable goods, such as furniture.

4.4 Elementary models of residential location

4.4.1 Introduction

The purpose of a residential-location model is to explain the distribution of people (grouped as families, or households) in the residences of a city or region. Clearly, whole complexes of mechanisms determine why people choose to live where they do. There is a demand side and a supply side: people demand residences in different locations, and constructors and developers supply various types of residence. Such a system forms what we usually call the housing market. It will be an imperfect market, of course: however, the essential task of a residential-location model will be to represent and explain the housing market in this way. On the demand side it is necessary to know the preference structures of different person types, and on the supply side it is necessary to know the goals, technology and managerial capabilities of developers, constructors, and related actors, as well as the impact of planners. In principle, it is then possible to begin to model the housing market. This remains the long-run goal. In the shorter run, however, the information which is needed to enable us to build true market models is unavailable and the traditional supply and demand curve concepts of economics do not lend themselves to theory building in *spatially disaggregated* systems. To date, a number of more elementary models have been developed. This section demonstrates the contribution of entropy-maximising methods to the understanding of these elementary models, and shows how improvements can be made by disaggregation.

4.4.2 Gravity models to allocate households around workplaces

The simplest hypothesis which can be made and developed is that households locate around workplaces, and, in particular, that the percentage of the people working in a particular location who live in a particular residential location falls as the distance (or, usually, travel cost) between the two locations increases. The most elementary such model is the residential component of the Lowry (1964) model. Let P_i be the residential population of zone i and let E_j be the number of jobs in zone j. Then Lowry's hypothesis is that

$$P_i = g\sum_j E_j \mathrm{f}(c_{ij}) \ , \tag{4.9}$$

where f is some decreasing function of the interzonal travel cost c_{ij}, and g is a constant calculated to ensure that P_i sums to a given total population P. That is, g is calculated by substituting from Equation (4.9) into

$$\sum_i P_i = P \ ; \tag{4.10}$$

so it can easily be seen that g is an inverse of an activity rate. Lowry, who was building a model of Pittsburgh, obtained the form of the function f from the Pittsburgh Transportation Study (Lowry, 1963).

This model is, of course, a very simple form of gravity model. In effect, it assumes that, if we define T_{ij} as the number of workers who live in i and work in j (so gT_{ij} is the number of people living in i in households whose workers work in j), then Lowry is really assuming that

$$T_{ij} = E_j \mathrm{f}(c_{ij}) \ . \tag{4.11}$$

Notice that with any model postulating some sort of spatial interaction—in this case involving a hypothesis that the interaction between residence and workplace determines residential location—the underlying assumptions can be illuminated if the *interaction* is used explicitly, as in Equation (4.11), rather than a total activity variable like T_{i*}—the P_i of Equation (4.9). This then facilitates further development of the model as Equation (4.11) exhibits it as a particularly simple form of gravity model. Before proceeding in this way, however, we can exhibit the constraints which would lead to Equation (4.11) on entropy-maximising principles. This will also facilitate further development.

To obtain Lowry's residential model, it is necessary to be particularly careful about the definition of entropy, and in this case it is convenient to construct a probability distribution by defining

$$p_{ij} = \frac{T_{ij}}{E_j} \tag{4.12}$$

rather than using T_{ij} directly as used above. (The two methods are equivalent—cf. Wilson, 1969a.) p_{ij} can be interpreted as the probability that a worker in zone j will live in zone i. Then the entropy of such a probability distribution can be taken as

$$S_j = -\sum_i p_{ij} \ln p_{ij} \ . \tag{4.13}$$

We would expect to maximise this S_j subject to the constraints

$$\sum_i p_{ij} = 1 \tag{4.14}$$

$$\sum_i p_{ij} c_{ij} = \bar{c}_j \ , \tag{4.15}$$

where $\bar{c}_j$ is the average expenditure on the journey to work for workers with jobs in j. Equation (4.15) is thus equivalent to the usual cost constraint stated for work trips to zone j only. However, the reader can easily check that if S_j is maximised subject to Equations (4.14) and (4.15), a constant of proportionality is introduced [as a result of Equation (4.14) as a constraint] which varies with j, and this is not the case in Lowry's model. Thus Equation (4.14) is dropped as a constraint. It should be replaced by a constraint of the form of Equation (4.10), but written in job terms; that is, E_j, summed over j, adds up to some given total employment. It is not possible to express this as a constraint on p_{ij} and still maximise S_j, as such a constraint would involve $p_{ij'}$'s for which $j' \neq j$. We can overcome this difficulty by replacing the definition of $p_{ij} = T_{ij}/E_j$ by $p_{ij} = T_{ij}/kE_j$, *not* imposing Equation (4.14) as a constraint, and then noting that k can be calculated so that all the E_j's sum to a given total. So maximise S_j subject to Equation (4.15) only, and this gives

$$p_{ij} = \exp(-\beta_j c_{ij}) \ , \tag{4.16}$$

where β_j is the Lagrangian multiplier associated with Equation (4.15). p_{ij} can be replaced by T_{ij}/kE_j, so Equation (4.16) becomes

$$T_{ij} = kE_j \exp(-\beta_j c_{ij}) \ . \tag{4.17}$$

If we then define

$$f_j(c_{ij}) = k \exp(-\beta_j c_{ij}) \ , \tag{4.18}$$

and further assume that β_j is independent of j, so that $f_j(c_{ij})$ can be written as $\mathrm{f}(c_{ij})$, then Equation (4.17) can be seen to be equivalent to

Equation (4.11) (and hence to Lowry's model). In this model [reverting now to Equation (4.11)] T_{ij} does not satisfy Equation (4.14) (not surprisingly, since it was dropped as a constraint): that is, $\sum_i T_{ij} \neq E_j$. So, if we maximise S_j subject to Equation (4.15) *and* Equation (4.14), and again assume that β_j is independent of j, we obtain

$$T_{ij} = B_j E_j \exp(-\beta c_{ij}) \tag{4.19}$$

$$B_j = \left[\sum_i \exp(-\beta c_{ij})\right]^{-1} , \tag{4.20}$$

and B_j is related to the Lagrangian multiplier associated with Equation (16).[8] The residential distribution p_i can always, of course, be obtained by summing in the usual way. Thus Equations (4.19) and (4.20) give the correct elementary residential model to represent the hypothesis which was originally stated: an allocation of workers to residences around workplaces governed by known travel-cost relationships.

In the model developed so far, the only attribute of residential areas considered in the model is accessibility to workplaces. The next stage is to use again the device introduced in Section 3.3 to include a term V_i^α into the model, where V_i in this case is some measure of the relative attractiveness of living in zone i. Equations (4.19) and (4.20) would then become

$$T_{ij} = B_j V_i^\alpha E_j \exp(-\beta c_{ij}) \tag{4.21}$$

and

$$B_j = \left[\sum_i V_i^\alpha \exp(-\beta c_{ij})\right]^{-1} . \tag{4.22}$$

If the V_i^α term represented the effect of scale economies, then it could be taken as the size of existing development in zone i. Since residential attractiveness is a complex notion (more complex than shopping attractiveness, for example), it would probably be better to construct some kind of composite index. There is a further reason why it is unsatisfactory to take V_i simply as the size of existing development: such a decision would have the effect of making 'fully developed' residential areas the most attractive. Some form of capacity-restraint mechanism would then have to be introduced into the model. This particular question will be discussed in Chapter 5 below, but it is clear that the use of 'size' as a proxy for 'attractiveness' leads to problems. These comments

[8] Of course, an entropy of the form of S obtained by summing additionally over j in Equation (4.13) could be maximised subject to the usual cost constraint [which is Equation (4.15), summed over j] and Equation (4.14); then it can easily be checked that the model given by Equations (4.19) and (4.20) is obtained directly. This approach has only been adopted to fit in with the derivation of the Lowry model which, as we have seen, does not easily lend itself to entropy-maximising assumptions!

also raise general doubts about the validity of such simple elementary models of residential location. Some of these doubts will be discussed more fully below and developments of these elementary models will be outlined. Note that the model given by Equations (4.21) and (4.22) is the equivalent for residential location of the Lakshmanan–Hansen shopping model.

One other point should be mentioned at this stage. Elementary models using the principles of the intervening-opportunities model can be developed. So as not to break the flow of the chapter, this is described in Appendix 3, along with other applications of the intervening-opportunities model.

4.4.3 A critical review

We can best see how to make progress by initially reviewing the elementary residential-location models presented above, and assessing how to meet the deficiencies.

Firstly, a general question which was raised in Section 4.1 above: if a constraint is dropped to develop an attraction-constrained gravity model as a residential-location model, as above, is this justifiable on behavioral grounds or is it merely an expression of ignorance? The question is particularly pointed in this case, as we have already seen in Chapter 2 that the usual transport journey-to-work model is doubly constrained (i.e. production–attraction-constrained). These two models, as presented, *are not compatible*. The residential-location model presented above obviously provides an estimate of T_{ij}, the journey to work, as well as T_{i*}, the distribution of residential activity; and such a T_{ij} will differ from that estimated by the doubly-constrained model of Chapter 2. These issues, then, will initiate our critical review. We shall go on to discuss the need to introduce income groups and house types as well as further disaggregation, and also the functioning of the supply side within elementary models. These points will be discussed in detail in subsequent sections, and in each case, appropriate developments of elementary models will be shown. Finally, we shall show how to integrate the various developments, at least in principle.

Firstly, then, is it justifiable to drop the constraint? The behavioral assumption implicit in doing so to develop an attraction-constrained model is that residential location is determined by the distribution of workplaces, and in particular by the workers' accessibility to workplaces. There are some obvious weaknesses in such an assumption: an increasing number of households, for example, of old people, contain no workers (a special case, which will be discussed further below). There are other sections of the population, such as those living in public housing where it would be more realistic to assume production-constrained locational behaviour, such people having more or less fixed residences and being allocated to workplaces around residences. Yet others, for example people migrating to a town from elsewhere, may be unconstrained in both location and job-choice decision. These three types of locational behaviour are explicitly recognised in a journey-to-work model developed by le Boulanger and Lissarrague (1966), and the two parameters which determine the proportions of the population in each group are estimated as parameters of a hybrid model. The actual models employed by le Boulanger and Lissarrague are not members of the family of spatial interaction models outlined in Chapter 3. They employ rather unusual procedures to estimate what we have previously called A_i and B_j.

We should also note that there is a fourth category of locational behaviour: the set of people who do not move and are not considering a move in a given time period. They, in a sense, are doubly constrained. This will provide the clue to making residential-location models compatible with transport journey-to-work models. It will also be noted that in introducing the fourth kind of locational behaviour it has proved necessary to introduce a time scale.

In Section 4.4.4 below, we shall set up a hybrid model to exhibit the four kinds of locational behaviour, and use the concept of time scale to make the model quasi-dynamic. In distinguishing the first three groups of locational behaviour (potential movement) from the last (nonmovement), pools of potential movers will be estimated within the model. This also improves the behavioral assumptions implicit in the model: it is often factors which are *internal* to the household (such as stage in life cycle, need for a bigger wage) which prompt the decision to move (either house or job or both, that is, to become a potential mover), and only then are *external* factors (such as attractiveness and accessibility, the variables of our models) taken into account in deciding where to move to. Thus, if the location model is preceded by a submodel to estimate potential movers, this corresponds to the likely two-stage decision process in the real world.

In Section 4.4.5, we shall remove the implicit assumptions to date that all people have the same incomes, and that all houses are the same type and cost the same irrespective of location. At the same time we shall take account of differing wage levels by location. An appropriate model will be exhibited with (Section 4.4.5) and without (Section 4.4.6) a simplifying assumption.

In Section 4.4.7, it will be noted that the model developments described in Section 4.4.5 facilitate the development of a model in which the supply-side's role is more explicit.

Further disaggregation will be outlined in Section 4.4.8. This development is especially concerned with social class variables (coarse indices of occupation) which ensure that appropriate people are allocated to appropriate jobs.

Finally, in Section 4.4.9, we show how the different model developments can be combined and how integrated models can be developed.

4.4.4 Four types of location behaviour; quasi dynamics

We have identified four types of locational behaviour. Let n be a person type index and let:

(a) $n = 1$ denote locationally unconstrained workers;
(b) $n = 2$ denote production-constrained workers (i.e. fixed residences);
(c) $n = 3$ denote attraction-constrained workers (i.e. fixed jobs);
(d) $n = 4$ denote production–attraction constrained workers (i.e. fixed residences and jobs, nonmovers)[9].

Let T_{ij}^n be the number of n-type workers living in i and working in j, c_{ij} be the cost of travel to work from i to j, and let g be an inverse activity rate, as before. P_i^n is the total n-type population in i and E_j^n the number of n-type jobs in j. Then

$$P_i^n = gT_{i*}^n \tag{4.23}$$

$$E_j^n = T_{*j}^n \,. \tag{4.24}$$

It has already been noted that a time scale is implicit in this formulation: the period during which $n = 4$ people are nonmovers. To fix ideas, let this period be one year, and let time t denote the beginning of the time period. t will not appear explicitly, and so the model at this stage will be quasi-dynamic only (on Hick's famous definition of a dynamic model). A t will be added in brackets to a variable if it is necessary to indicate to which time period it refers. Thus $E_j^n(t)$ would be the number of n-type jobs in zone j in a period of one year from time t.

As a basis for building a hybrid model, we would expect to be given the following information for each of the four groups:

(a) $n = 1$: P_*^1 (and $gE_*^1 = P_*^1$ of course);
(b) $n = 2$: P_i^2;
(c) $n = 3$: E_j^3;
(d) $n = 4$: P_i^4 and E_j^4;

where an asterisk is being employed in the usual way to indicate summation. For consistency, we must have

$$P_*^4 = gE_*^4 \,. \tag{4.25}$$

There are various ways in which P_*^1, P_i^2, E_j^3, P_i^4, and E_j^4 can be estimated outside the location model itself. Such a submodel represents the first

[9] We are using the terminology introduced in Chapter 3. People in the fourth category will be referred to below as 'nonmovers'. This is a convenient shorthand. It does not mean that people in this location group never move, but, if they move, it is in such a way that their assignment to jobs and residences remains fixed. The best way to interpret this assumption is that people in the fourth category are locationally fixed, but their journey-to-work pattern satisfies a gravity model: making the assumption in this way guarantees this [as appears in Equation (4.55) below]. The alternative, and in some ways more obvious solution, would be to take T_{ij}^4 as given and fixed. The reader can easily adjust the argument of this section to build a model with this form of assumption if required.

part of the two-stage decision process mentioned earlier. For the purposes of this chapter it is only necessary to illustrate that reasonable schemes can be set down for estimating these quantities. One such scheme follows.

Let X^k be a set of variables which between them determine new household formation and net immigration for the study area. Then P_*^1 could be estimated by regression analysis using an equation of the form

$$P_*^1 = \sum_k a_1^k X^k , \tag{4.26}$$

where a_1^k is a set of coefficients.

The remaining quantities in the list have to be modelled *for each zone.* Let Y_i^k and Z_i^k be sets of variables which describe propensity to be residence-fixed and propensity to be job-fixed in zone i. The total number of people in zone i with fixed residences is then $P_i^2+P_i^4$, and the total number of jobs in zone j with people fixed in jobs is $E_j^3+E_j^4$. So appropriate regression equations are

$$P_i^2+P_i^4 = \sum_k a_{2/4}^k Y_i^k \tag{4.27}$$

$$E_j^3+E_j^4 = \sum_k a_{3/4}^k Z_j^k , \tag{4.28}$$

where $a_{2/4}^k$ and $a_{3/4}^k$ are suitable sets of coefficients. For example, Y_i^k may include the number of council houses in i and variables describing the occupation/age mix of the population, and Z_j^k may include variables describing the occupation mix. The coefficients, in this formulation, are dismobility coefficients, but are obviously related to rates of movement. Generally speaking, we would expect the model to have a quasi-dynamic recursive form, as explained earlier, and this would probably be achieved by taking values of X, Y, and Z referring to a time period earlier than that for the dependent variables, thus building lags into the model. It can be left as an exercise to the reader to make this explicit.

It is still necessary to define further coefficients to separate fully the groups (2), (3), and (4). Let

$$P_i^2 = yP_i^4 \tag{4.29}$$

and

$$E_j^3 = zE_j^4 . \tag{4.30}$$

Note that y and z must be related through the consistency condition, Equation (4.25). This can be made explicit as follows: Equations (4.27)–(4.30) can each be summed over spatial indices to give

$$P_*^2+P_*^4 = \sum_k a_{2/4}^k Y_*^k \tag{4.31}$$

$$E_*^3+E_*^4 = \sum_k a_{3/4}^k Z_*^k \tag{4.32}$$

$$P_*^2 = yP_*^4 \tag{4.33}$$

$$E_*^3 = zE_*^4 . \tag{4.34}$$

P_*^2, E_*^3, P_*^4, and E_*^4 can be eliminated from Equations (4.25) and (4.31)–(4.34) to give

$$\frac{(1+z)\sum_k a_{2/4}^k Y_*^k}{(1+y)\sum_k a_{3/4}^k Z_*^k} = g \tag{4.35}$$

as the relation between y and z.

An alternative, and perhaps simpler scheme, for identifying potential movers would be to take the total population and job distributions from the previous time period, disaggregated, for example, by social class (n') and occupation (n'') and to express P_i^2, E_j^3, P_i^4, and E_j^4 as functions of that population. P_*^1 would be estimated as before. Thus, take $P_i^{*n'}(t-1)$ and $E_j^{*n''}(t-1)$ as the total population of social class n' in zone i at $t-1$ (at either the beginning or end of the period, depending on the extent to which a lag is thought desirable) and the total number of jobs for occupation group n'' at $t-1$) (or use n'' to denote industry). Then we could define $y_2^{n'}$, $y_4^{n'}$, $z_3^{n''}$, and $z_4^{n''}$ as coefficients such that

$$P_i^2 = \sum_{n'} y_2^{n'} P_i^{*n'}(t-1) \tag{4.36}$$

$$E_j^3 = \sum_{n''} z_3^{n''} E_j^{*n''}(t-1) \tag{4.37}$$

$$P_i^4 = \sum_{n'} y_4^{n'} P_i^{*n'}(t-1) \tag{4.38}$$

$$E_j^4 = \sum_{n''} z_4^{n''} E_j^{*n''}(t-1) \tag{4.39}$$

where, again, the coefficients can be estimated by regression analysis. It may well be that a model of this second form would be easier to estimate than that presented earlier, as there are no interrelations between the coefficients. There is one estimation problem, however, which must be discussed in either case.

In order to use regression analysis, or some analogous technique, values of the dependent variables on the left-hand sides of the equations need to be known. How can we measure P_i^2, E_j^3, P_i^4, and E_j^4? The question arises because P_i^2 and E_j^3 are *potential* movers: that is, they include people who think about changing residence or job but do not in fact do so. We can only measure groups of *actual* movers. Such a difficulty can only be overcome by making an assumption and testing it. For example, the simplest possible assumption is that the ratio of nonmovers (out of the potential movers) to potential movers is a constant. Such a constant could be estimated to give a best fit in the location model.

We can now assume that a procedure can be given to estimate P_*^1, P_i^2, E_j^3, P_i^4, and E_j^4, and so our interaction variable T_{ij}^n must satisfy the following groups of constraints:

(a) $n = 1$:

$$g\sum_i \sum_j T_{ij}^1 = P_*^1 \tag{4.40}$$

$$\sum_i \sum_j T_{ij}^1 c_{ij} = C^1; \tag{4.41}$$

(b) $n = 2$:

$$g\sum_j T_{ij}^2 = P_i^2 \tag{4.42}$$

$$\sum_i \sum_j T_{ij}^2 c_{ij} = C^2; \tag{4.43}$$

(c) $n = 3$:

$$\sum_i T_{ij}^3 = E_j^3 \tag{4.44}$$

$$\sum_i \sum_j T_{ij}^3 c_{ij} = C^3; \tag{4.45}$$

(d) $n = 4$:

$$g\sum_j T_{ij}^4 = P_i^4 \tag{4.46}$$

$$\sum_i T_{ij}^4 = E_j^4 \tag{4.47}$$

$$\sum_i \sum_j T_{ij}^4 c_{ij} = C^4; \tag{4.48}$$

where C^n is the total expenditure on the journey to work by persons of type n (i.e. in the nth locational behaviour group). As we know from Chapter 2 experience, admitting a C^n for each group, will leave open the possibility of a different coefficient in the transport-cost deterrence function for each person type. We can also use the rules of Section 3.3 of Chapter 3 to introduce, where appropriate, terms $V_i^{\alpha_1}$ and $W_j^{\alpha_2}$ to measure, respectively, the relative attraction of residential zone i and of workplace zone j. The problem of measuring V_i was discussed in Section 4.4.2 above. For W_j, something like the local wage rate, or a proxy for it, could be used. The entropy-maximising model which results from the set of constraint Equations (4.40)–(4.48) can be expressed in groups of equations:

(a) $n = 1$:

$$T_{ij}^1 = KV_i^{\alpha_1}W_j^{\alpha_2}\exp(-\beta^1 c_{ij}), \tag{4.49}$$

where

$$K = P_*^1\left[\sum_i \sum_j V_i^{\alpha_1}W_j^{\alpha_2}\exp(-\beta^1 c_{ij})\right]^{-1}; \tag{4.50}$$

(b) $n = 2$:

$$T_{ij}^2 = A_i^2 P_i^2 W_j^{\alpha_2}\exp(-\beta^2 c_{ij}), \tag{4.51}$$

where

$$A_i^2 = \left[\sum_j W_j^{\alpha_2}\exp(-\beta^2 c_{ij})\right]^{-1}; \tag{4.52}$$

(c) $n = 3$: $$T_{ij}^3 = B_j^3 V_i^{\alpha_1} E_j^3 \exp(-\beta^3 c_{ij}) \,, \tag{4.53}$$

where $$B_j^3 = \left[\sum_i V_i^{\alpha_1} \exp(-\beta^3 c_{ij})\right]^{-1} \;; \tag{4.54}$$

(d) $n = 4$: $$T_{ij}^4 = A_i^4 B_j^4 P_i^4 E_j^4 \exp(-\beta^4 c_{ij}) \,, \tag{4.55}$$

where $$A_i^4 = \left[\sum_j B_j^4 E_j^4 \exp(-\beta^4 c_{ij})\right]^{-1} \tag{4.56}$$

and $$B_j^4 = \left[\sum_i A_i^4 P_i^4 \exp(-\beta^4 c_{ij})\right]^{-1} \,. \tag{4.57}$$

This is a genuine hybrid model (that is, the different parts are closely interlinked) because the measures of attractiveness, V_i and W_j, are common to three of the four parts of the model, and each of these will depend ultimately on the distributions of all residences and employment, formed by summing over the four parts of the model. Further, the parts are interlinked because P_*^1, P_i^2, E_j^3, P_i^4, and E_j^4 are all related.

The whole of this section constitutes a rather lengthy answer to the initial question of whether we are justified in dropping a constraint to obtain a location model. We have seen that location models can be developed, but that we should recognise the different possible kinds of locational behaviour and find out empirically which sectors of the population fall into which group. We have also seen that this forced us to develop a quasi-dynamic recursive form of model (at least implicitly, and which is easy to make explicit).

Finally, we can solve the problem of relating the residential-location model to the transport journey-to-work model. Suppose the implicit time period of the quasi-dynamic location model is very short. Then most people will be in group (d), and the resulting model (within the residential model) of the journey to work will be pretty much a production–attraction-constrained model. This indicates the answer to the conundrum: if a journey to work is required *at an instant of time*, given distributions of population and employment for that time, then a production–attraction-constrained model should be used[(10)]. In a forecasting situation, however, the recursive hybrid residential-location model, or something like it, should be used. Thus, in urban transportation studies, it is reasonable to use the production–attraction-constrained model for a journey-to-work forecast in circumstances where the residential location forecast is exogenous to the model. This work does indicate, however, that a critical review of procedures for making residential-location forecasts in urban studies is called for.

An advantage of using the recursive model for growth forecasting and

[(10)] So this is another justification for the definition of category (d) of locational behaviour, as discussed in footnote (9).

the doubly-constrained model for an 'instant' analysis of a transportation pattern is that different time lags can be incorporated to account for the effects of transportation network changes (in effect, changes in the c_{ij}'s and hence accessibilities). For example, in the transportation model, changes of destination and route can be considered to take place almost without a lag, while in the residential-location model, quite long lags can be introduced.

4.4.5 Disaggregation, with a simplifying assumption

We saw in the review, Section 4.4.3, that we need to disaggregate elementary residential-location models to allow for at least the following:
(a) different income groups;
(b) different wage levels by location;
(c) different types of house;
(d) variation in the price of houses by location.

This disaggregation can be tackled in two stages, firstly, making a simplifying assumption to illustrate the general principles. The major simplifying assumption, which will be relaxed in the next section, is that each household has one, and only one, worker. With this assumption, we can concentrate on assigning workers to jobs and residences without having to tackle the further complexities of household structure. Our first task will be to disaggregate the Lakshmanan–Hansen type of model of Section 4.4.2 (although it will be convenient to model a production-attraction-constrained situation) and to leave the further developments of Section 4.4.4, models incorporating different types of locational behaviour and a quasi-dynamic recursive framework, until Section 4.4.9, where we explore the integration of the various developments.

Disaggregation can be achieved by defining T_{ij}^{kw} to be the number of workers who live in zone i in a house of type k, and who work in zone j earning wage w. (We shall assume that earnings are the only source of income.) Let H_i^k be the number of houses in zone i of type k, and let E_j^w be the number of jobs in zone j offering wage w. k will be a size/age/condition index. c_{ij} will be the cost of travel from i to j, as usual. Then the interaction variable T_{ij}^{kw} must satisfy the following constraints:

$$\sum_j \sum_w T_{ij}^{kw} = H_i^k \tag{4.58}$$

$$\sum_i \sum_k T_{ij}^{kw} = E_j^w \tag{4.59}$$

$$\sum_i \sum_j \sum_k T_{ij}^{kw} c_{ij} = C^w , \tag{4.60}$$

where we assume that house types and wage levels are represented in discrete groups. It is convenient to continue to use the discrete formalism, though continuous variables could be used if necessary.

Integral signs would then replace summation signs in equations involving sums of those variables. The constraint Equations (4.58)–(4.60) will give rise to a set of doubly-constrained interlinked gravity models of a form we could anticipate from previous experiences. Note that, by introducing C^w, the transport parameter in the model will take the form β^w, making trip lengths a function of income rather than person type n; we shall retain this whenever income appears explicitly. Note also that, by introducing H_i^k and E_j^w as exogenous information required for this model, we are beginning to represent the supply side: the values of these variables will be determined by consortia of suppliers, planners, developers, employers, and so on. Such variables would be endogenous to a more general model, and this point will be discussed further in Section 4.4.7 below.

We need one further constraint equation. It is necessary to find a way of expressing the fact that workers will, within limits, live in houses which they can afford. Let us define p_i^k as the price of a type k house in zone i, and q^w as the average percentage of income (after transport costs have been deducted) which a member of income group w spends on housing. The problem is to relate these two variables. We saw previously that cost constraints of the form of Equation (4.60) lead to negative exponential distributions being introduced into the model. This time we want to incorporate a distribution which shows the spread of actual expenditure on housing around the mean and relates this to the house price variable p_i^k. Since we want to compare this expenditure with the actual *money* cost of the journey to work, let c'_{ij} be that component of the usual *generalised* journey-to-work cost c_{ij}, which is the actual money paid. The simplest assumption which could be made is that such a distribution is normal. This can be introduced by adding the constraint

$$\sum_i \sum_j \sum_k T_{ij}^{kw}[p_i^k - q^w(w - c'_{ij})]^2 = \sigma^{w^2} \tag{4.61}$$

where σ^{w^2} will be the variance of the normal distribution for income group w. We can maximise entropy in the form

$$S = -\sum_i \sum_j \sum_k \sum_w \ln T_{ij}^{kw}! \tag{4.62}$$

subject to the constraint Equations (4.58)–(4.61) by forming the Lagrangian

$$\begin{aligned} \mathcal{L} = &-\sum_i \sum_j \sum_k \sum_w \ln T_{ij}^{kw}! - \sum_i \sum_k \lambda_i^{(1)k}\Big(\sum_j \sum_w T_{ij}^{kw} - H_i^k\Big) \\ &- \sum_j \sum_w \lambda_j^{(2)w}\Big(\sum_i \sum_k T_{ij}^{kw} - E_j^w\Big) - \sum_w \beta^w\Big(\sum_i \sum_j \sum_k T_{ij}^{kw} c_{ij} - C^w\Big) \\ &- \sum_w \mu^w \Big\{\sum_i \sum_j \sum_k [p_i^k - q^w(w - c'_{ij})]^2 - \sigma^{w^2}\Big\} , \end{aligned} \tag{4.63}$$

where $\lambda_i^{(1)k}$, $\lambda_j^{(2)w}$, β^w, and μ^w are the Lagrangian multipliers associated with Equations (4.58)–(4.61), respectively, and solving

$$\frac{\partial \mathcal{L}}{\partial T_{ij}^{kw}} = 0 \tag{4.64}$$

and the constraint equations. This gives

$$T_{ij}^{kw} = A_i^k B_j^w H_i^k E_j^w \exp(-\beta^w c_{ij}) \exp\{-\mu^w [p_i^k - q^w(w - c'_{ij})]^2\}, \tag{4.65}$$

where

$$A_i^k = \frac{\exp(-\lambda_i^{(1)k})}{H_i^k} \tag{4.66}$$

and

$$B_j^w = \frac{\exp(-\lambda_j^{(2)w})}{E_j^w} \tag{4.67}$$

We can obtain A_i^k and B_j^w explicitly by using Equations (4.58) and (4.59) to give

$$A_i^k = \left(\sum_j \sum_w B_j^w E_j^w \exp(-\beta^w c_{ij}) \exp\{-\mu^w [p_i^k - q^w(w - c'_{ij})]^2\}\right)^{-1} \tag{4.68}$$

and

$$B_j^w = \left(\sum_i \sum_k A_i^k H_i^k \exp(-\beta^w c_{ij}) \exp\{-\mu^w [p_i^k - q^w(w - c'_{ij})]^2\}\right)^{-1}. \tag{4.69}$$

It is obviously necessary to be careful to ensure that all costs and prices relate to the same time period.

If the last term of the product forming the right-hand side of Equation (4.65) were omitted, then the model would simply be a disaggregated doubly-constrained residential-location cum journey-to-work model. In other words, it is not as unfamiliar as might appear at first sight. The final term makes the disaggregation worthwhile and effective: if $q^w(w - c'_{ij})$, the average available housing expenditure, is very different (either way) from p_i^k, then very few wage w people, working in j, would be allocated to a type k house in i. It is obviously the sort of device we need to build appropriate income effects into a residential model. For example, if in a large city the only cheap accommodation which is available is old accommodation in the city centre, or newer accommodation in the distant suburbs, then low-income households will be allocated predominantly to this accommodation. In a purely spatial interaction kind of model, this would not happen, and could lead to misleading results which may look all right in aggregate, but which may hide some severe planning problems.

The data problems associated with calibrating a model of this type are clearly immense. However, they are not intractable: on the one hand, the development of this type of model should encourage the collection of appropriate data; on the other, even in the short run, proxy variables can be constructed for indices such as house type.

4.4.6 Removal of the simplifying assumption

There are at least two ways of removing the major simplifying assumption of the previous section that there is one, and only one, worker per household. Theoretical problems do arise, however, in either case. The alternatives arise because of two possible assumptions: that only the income of the head of the household is relevant to the residential-location decision, or that all household income is relevant. 'Intermediate' assumptions are also possible.

To tackle the first assumption, it is necessary to distinguish heads of households. Developing the notation of the previous section, let $T_{ij}^{kwn'}$ be the number of workers living in zone i in a house of type k, working in j for wage w, and let $n' = 1$ indicate headship of household, $n' = 0$ not (i.e. a 'dependent' worker). Let r be the mean number of workers in households. Then the Equation (4.58) can be written in the form

$$\sum_j \sum_w \sum_{n'} \frac{T_{ij}^{kwn'}}{r} = H_i^k \ , \tag{4.70}$$

or, alternatively, in the form

$$\sum_j \sum_w \sum_{n'} \delta_{n'1} T_{ij}^{kwn'} = H_i^k \ , \tag{4.71}$$

which are each ways of expressing the condition that houses are assigned to *heads* of households. All workers are assigned to jobs, so Equation (4.59) becomes

$$\sum_i \sum_k \sum_{n'} T_{ij}^{kwn'} = E_j^w \ , \tag{4.72}$$

and all pay for the journey to work, so Equation (4.60) becomes

$$\sum_i \sum_j \sum_k \sum_{n'} T_{ij}^{kwn'} c_{ij} = C^w \ . \tag{4.73}$$

The equivalent of Equation (4.61) is

$$\sum_i \sum_j \sum_k \sum_{n'} T_{ij}^{kwn'} \delta_{n'1} \left[p_i^k - q^w(w - c_{ij}')\right]^2 = \sigma^{w^2} \tag{4.74}$$

where the $\delta_{n'1}$ term ensures that only the head of the household contributes to housing costs.

In order to derive a specific form of entropy-maximising model, we must decide which of Equations (4.70) and (4.71) to use. It seems better on balance to choose Equation (4.70), which will ensure that 'dependent' workers form a constant proportion of the total in each zone. Using Equation (4.71), they would be unconstrained. Thus, maximising

$$S = -\sum \ln T_{ij}^{kwn'}! \tag{4.75}$$

subject to Equations (4.70) and (4.72)–(4.74) gives

$$T_{ij}^{kw1} = A_i^k B_j^w r H_i^k E_j^w \exp(-\beta^w c_{ij}) \exp\{-\mu^w [p_i^k - q^w(w - c'_{ij})]^2\} \tag{4.76}$$

for $n' = 1$, and

$$T_{ij}^{kw0} = A_i^k B_j^w r H_i^k E_j^w \exp(-\beta^w c_{ij}) \tag{4.77}$$

for $n' = 0$, and where

$$A_i^k = \frac{\exp(-\lambda_i^{(1)k})}{rH_i^k} \tag{4.78}$$

and

$$B_j^w = \frac{\exp(-\lambda_j^{(2)w})}{E_j^w}, \tag{4.79}$$

and where $\lambda_i^{(1)k}$, $\lambda_j^{(2)w}$, β^w, and μ^w are the Lagrangian multipliers associated with Equations (4.70) and (4.72)–(4.74), respectively. A_i^k and B_j^w can be obtained straightforwardly by substituting from Equations (4.76) and (4.77) into Equations (4.70) and (4.72) and this is left as an exercise to the reader. β^w and μ^w can be obtained by the usual calibration process.

We can now turn to the alternative hypothesis: that all household income contributes to housing costs. The problem to be tackled here is that of systematically grouping individual wages w into *household* incomes I for households with more than one worker. The n' superscript can be dropped, and we can revert to T_{ij}^{kw} as the number of workers living in i in a house of type k, and working in zone j for wage w. We now postulate that we know, or can find, a distribution $\pi(w \mid I)$, the probability of a w wage earner being in a household of income I. The constraint Equations (4.70) and (4.72)–(4.74) can now be amended in an obvious way to

$$\sum_j \sum_w \frac{T_{ij}^{kw}}{r} = H_i^k \tag{4.80}$$

$$\sum_i \sum_k T_{ij}^{kw} = E_j^w \tag{4.81}$$

$$\sum_i \sum_j \sum_k T_{ij}^{kw} = C^w \tag{4.82}$$

$$\sum_i \sum_j \sum_k \frac{\pi(w \mid I) T_{ij}^{kw}}{r} [p_i^k - q^I(I - c'_{ij})]^2 = \sigma_{wI}^2 . \tag{4.83}$$

Only the last of the four equations needs much explanation. $\pi(w \mid I)T_{ij}^{kw}$ is the number of w wage earners living in households of income I, and so the Equation (4.83) implies that such a worker will locate according to his household income. It is a slight weakness of this formulation that the individual's transport costs appear in Equation (4.83) and not either all transport costs incurred by the household, or the transport cost of the head of the household. Neither of these quantities can be identified in this formulation of the model. The implication of Equation (4.83) as stated is a reasonable approximation, however. If I is large, large scale commuting for all members of the family becomes feasible, otherwise not. The resulting entropy-maximising model can be written as

$$T_{ij}^{kw} = A_i^k B_j^w H_i^k E_j^w \exp(-\beta^w c_{ij}) \exp\left\{-\mu^{wI} \frac{\pi(w|I)}{r} [p_i^k - q^I(I - c'_{ij})]^2\right\} \quad (4.84)$$

where β^w and μ^{wI} are Lagrangian multipliers, A_i^k and B_j^w are related to Lagrangian multipliers, and can all be given values in the usual ways.

4.4.7 Modelling the supply side

The residential-location models presented so far are all essentially demand-side models: they locate the consumers of residences, on the basis of indices of attraction at the residential end of the work trip, or *on the basis of given distributions of houses of different types, H_i^k*. This is the first glimmer of how to cope with the supply side, and to extend the elementary models so that they begin to look like market models, with interacting demand and supply.

To fix ideas, let us reintroduce the person-type index n to represent kinds of locational behaviour, but let us assume one, and only one, worker per household, and only two types of locational behaviour: potential residence movers with fixed workplaces ($n = 1$) and the nonmovers[11] ($n = 2$). This will help us to construct a supply-side model and will also make a path towards integrating previous developments, albeit in simplified form. The spatial interaction variable is T_{ij}^{kwn}, and it satisfies

$$\sum_j \sum_w T_{ij}^{kw1}(t) = \Delta H_i^k(t) \quad (4.85)$$

$$\sum_j \sum_w T_{ij}^{kw2}(t) = H_i^k(t) \quad (4.86)$$

$$\sum_i \sum_k \sum_n T_{ij}^{kwn}(t) = E_j^w(t) \quad (4.87)$$

$$\sum_i \sum_j \sum_k \sum_n T_{ij}^{kwn}(t) c_{ij}(t) = C^w(t) \quad (4.88)$$

$$\sum_i \sum_j \sum_k T_{ij}^{kw1}(t)\{p_i^k(t) - q^w(t)[w(t) - c'_{ij}(t)]\}^2 = \sigma_w(t) \quad (4.89)$$

where all the variables are obvious extensions of previously defined variables, except ΔH_i^k and H_i^k. ΔH_i^k is the number of houses made available by the

(11) See footnote (9); the same point applies here. The assumption is perhaps less satisfactory in this case, given both house-type and income-group disaggregation. The reader can easily develop an alternative model, if required, based on the alternative assumption that $T_{ij}^{kw2}(t)$ is given [or can be calculated from $T_{ij}^{kw2}(t-1)$].

supply side to potential movers; H_i^k is the number of houses occupied by nonmovers. To clarify the quasi-dynamic nature of the model, an index t has been added to indicate the start of the time period to which a variable refers. Note that the constraint Equation (4.89) is only applied to the potential movers ($n = 1$).

The maximum-entropy model which results from this set of constraints can be written down as follows:

$$\text{(a)}\quad n = 1: \qquad T_{ij}^{kw1}(t) = A_i^{k1}(t)B_j^w(t)\Delta H_i^k(t)E_j^w(t)\exp[-\beta^w c_{ij}(t)] \times \exp[\![-\mu^w\{p_i^k(t)-q^w(t)[w(t)-c'_{ij}(t)]\}^2]\!] \; ; \tag{4.90}$$

$$\text{(b)}\quad n = 2: \qquad T_{ij}^{kw2}(t) = A_i^{k2}(t)B_j^w(t)H_i^k(t)E_j^w(t)\exp[-\beta^w c_{ij}(t)] \; ; \tag{4.91}$$

where β^w and μ^w are the Lagrangian multipliers associated with Equations (4.88) and (4.89), and A_i^{k1}, A_i^{k2}, and B_j^w are derived from the Lagrangian multipliers associated with Equations (4.85)-(4.87), respectively. These quantities can all be given values in the usual way.

The total stock of houses by type and location is assumed known. It can also be assumed that T_{i*}^{k*2}, the nonmovers, and $T_{*j}^{*w1}(t)$, the potential movers, can be estimated by some procedure (for example, as suggested in Section 4.4.4) from $T_{ij}^{kw*}(t-1)$. Because of our simplifying assumption, we must have

$$H_i^k(t) = T_{i*}^{k*2}(t) \, . \tag{4.92}$$

ΔH_i^k, then, is partly supplied by potential movers, and partly by new houses coming on to the market; some new houses may be the result of redevelopment involving the demolition of old ones. Let $\Delta H_i^k(1)$ be the set of houses made available by potential movers thinking of leaving houses they are already in [which can be calculated from $H_i^k(t)$ and a knowledge of the existing total stock]; let $\Delta H_i^k(2)$ be the set of houses remaining to be supplied on vacant land; and let $\Delta H_i^k(3)$ be the set of houses supplied through redevelopment. It is assumed, of course, that the residents of such houses join the potential mover pool. A supply-side part of the model should now estimate $\Delta H_i^k(2)$ and $\Delta H_i^k(3)$. The interaction of the demand and supply sides should also determine the prices p_i^k in the form of building costs h^k and a location premium for each house type l_i^k, so

$$p_i^k = h^k + l_i^k \, . \tag{4.93}$$

We can assume, without too much loss of generality, that h^k is known and given. Also, for simplicity, let us assume for the time being that there is no redevelopment.

We can now begin to see how to simulate market behaviour under different sets of rules. It can be seen at the outset that with this type of model, it is unreasonable to try to produce unique 'forecasts' of the development of the market. In particular, zones with vacant land, different developers, working partly intuitively, would use the same land to build different types of houses at different prices. The model being developed here will forecast the impact of sets of decisions of that kind, and can be used to produce or forecast the 'best' overall plan, or working of the market, only if appropriate evaluative welfare criteria can be established. There are at least two ways to explore this problem: (a) postulating an essentially public-welfare-maximising system, (b) postulating an essentially private-profit-maximising system. We shall explore each in turn.

Good public planning would only be possible, given a knowledge of the preference structure of individuals in the community with respect to housing and the journey to work relative to incomes. Such knowledge might be expressed by asserting that, within an income group w, there was a 'need' for different types of housing combined with various permissible journey-to-work lengths. Thus $P^w(k, c)$ might be the proportion of income group w 'needing' a house of type k and a journey to work of cost c. Such a function could be used, in other words, to represent our knowledge of the preference structure referred to earlier. Note that

$$P^w(k, c) = \sum_{\substack{i, j \\ \text{such that} \\ c_{ij} = c \pm \delta}} T_{ij}^{kw*} \Big/ T_{**}^{**w*} , \tag{4.94}$$

where δ is a constant. A planned set of ΔH_i^k's will lead to a prediction of T_{ij}^{kw*}, so the public planning task can now be seen as identifying the sequence of ΔH_i^k's in each time period, so that $P^w(k, c)$ as predicted by the model (hopefully reproducing what happens) tends to the preferred value of $P^w(k, c)$. This is obviously a complex optimisation process and will not be pursued further here. Heuristic methods are obviously possible at the very least.

The private-profit-maximising situation is a different one. We can assume that there are enough developers, each trying to maximise his own profits, and that no single one can dominate the market. Any one developer will then assess what price he could charge for different types of houses on plots of land which he owns or can buy. If the market is sufficiently competitive, he will not be able to make 'excess' profits. He will, however, try to identify that section of the market which gives him the biggest profit per unit of land in a zone. It is easy to see how the model will help him to do this, again at least on a trial basis. It is also easy to see how the private 'solution' will be different from the public one, as the most potentially profitable parts of the market could be overdeveloped. This is analogous to the 'ice-cream men on a linear beach' situation (Hotelling, 1929, explained by Wilson, 1968).

So, we begin to see how to model the supply side. There are obviously many situations between the two extreme examples considered here which could be studied in greater depth.

A final point of interest should be noted in this section: the concept of potential movers is useful in that it allows older (and probably richer) families to move into *new* houses, while new households or immigrants will have to move into the houses vacated by them. This effect is commonly observed in the real world, and it is useful to have it represented in the model.

4.4.8 Further disaggregation

Consider the spatial interaction variable T_{ij}^{kwn}, already defined. n is a person-type index and has been used to indicate type of locational behaviour. We have also, in other circumstances, used n to indicate headship of household or not. We can consider the person-type index n to be a composite index

$$n = nn'n''n''' \ldots, \tag{4.95}$$

where we might have n as the type of locational behaviour, n' the headship of household index, n'' the social class index, n''' the car availability index, and so on. In models such as the ones described, these further subdivisions can be understood: it can be assumed that there is summation over any indices not shown explicitly.

Obviously, some further disaggregation is desirable. We have already mentioned the need for a social-class/occupation index to ensure that people are allocated to the right kind of jobs. Even a coarse blue-collar/white-collar split may be useful in practice. It is also necessary to disaggregate by transport mode, as already indicated for the shopping model of Section 4.3. Again, this is left as an exercise for the reader.

One general point can be noted in this context. So far, it has been assumed that all indices refer to discrete groups. In some cases, income w is perhaps an example, it may be more appropriate to assume continuous variation, and to replace summation signs by integral signs for appropriate variables. This is relevant to the discussion of disaggregation as follows: in a spatial model, it only needs a relatively small number of 'cells' to give problems. Thus, with the variable T_{ij}^{kwn}, if there are 50 zones, 5 house types, 5 income groups, and 4 types of locational behaviour, there are $50^2 \times 5 \times 5 \times 4 = 250000$ cells or population categories. Few sample surveys would provide enough observations in each cell of such an array to make model calibration possible. There are various ways out of such a dilemma: calibration can use aggregates of the array, such as T_{*j}^{kw*} for example, or a trip-length distribution; the fact that many cell entries will be zero can be recognised and calibration can use nonzero entries only, provided the zeros can be systematically identified; finally, some indices, such as income, can be replaced by continuous distributions described by one or two parameters, so that only these parameters have to be estimated.

One particular point should be mentioned here: we have implicitly assumed so far that each household has at least one worker. Some do not, and their locational behaviour changes accordingly. There are two ways of handling the situation by disaggregation: to introduce a fifth locational behaviour group, or to introduce an additional 'social class'. This class would receive appropriate incomes, and the c_{ij} matrix for this class should be modified so that all the c_{ii}'s are zero, and the other elements infinite. The income would then be received at home.

4.4.9 Integrated models

The structure of the discussion in Section 4.4 to date is shown in Figure 4.1.

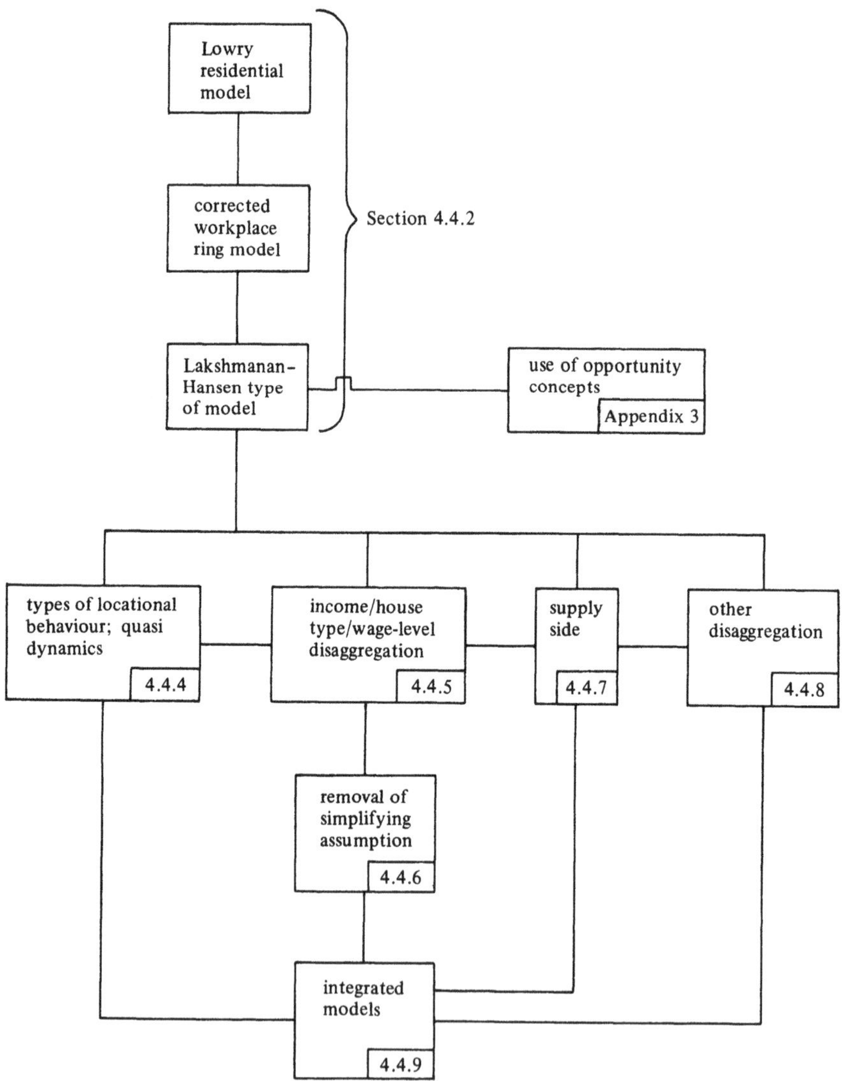

Figure 4.1. Structure of discussion of elementary residential-location models.

The purpose of this section is to call attention to the fact that the different individual developments of elementary models can be integrated. One simple example of such integration was given in Section 4.4.7, where a model was developed with two types of locational behaviour and the income/house-type/wage-level disaggregation. Only the general principles of integration will be discussed here; the details will be left as exercises for the reader.

As an illustration, let us find a way of combining the four kinds of locational behaviour with the income/house-type/wage-level disaggregation. This task is simply to introduce the four types of locational behaviour in place of the two in the integrated model of Section 4.4.7. Let ΔH_i^k be the number of houses available to potential house movers ($n = 1$ and 3), and let ΔE_j^w be the jobs available to potential job movers ($n = 1$ and 2). In the latter case, let us make a distinction between ΔE_j^{w1}, the jobs available to $n = 1$ people (who will generally be in new or immigrant households, commanding only low incomes) and ΔE_j^{w2}, available to $n = 2$ people (who will generally be people looking for better, and therefore higher-income, jobs). Then

$$\Delta E_j^w = \Delta E_j^{w1} + \Delta E_j^{w2}, \tag{4.96}$$

and an income distribution identical with that for E_j^w (the fixed jobs) can be assumed for ΔE_j^w, one with a lower mean for ΔE_j^{w1}, and therefore one with a higher mean for ΔE_j^{w2}. ΔH_i^k, ΔE_j^{w1}, and ΔE_j^{w2} will each consist partly of new units (so on the job side we are beginning to take account of changing industrial mix in effect) and partly of units made available by potential movers, as described for housing in Section 4.4.7. If we let r_n be the mean size of a type n household, and let n' be a headship index, as before, then we can write down a set of constraints which relaxes the assumption of only one worker per household. The constraints are

$$\sum_j \sum_w \sum_{n=1,3} \frac{T_{ij}^{kwn}}{r^n} = \Delta H_i^k \tag{4.97}$$

$$\sum_j \sum_w \sum_{n=2,4} \frac{T_{ij}^{kwn}}{r^n} = H_i^k \tag{4.98}$$

$$\sum_i \sum_k T_{ij}^{kw1} = \Delta E_j^{w1} \tag{4.99}$$

$$\sum_i \sum_k T_{ij}^{kw2} = \Delta E_j^{w2} \tag{4.100}$$

$$\sum_i \sum_k \sum_{n=3,4} T_{ij}^{kwn} = E_j^w \tag{4.101}$$

$$\sum_i \sum_j \sum_k \sum_n T_{ij}^{kwn} c_{ij} = C^w \tag{4.102}$$

$$\sum_i \sum_j \sum_k \sum_{n=1,3} T_{ij}^{kwn} \delta_{n'1} [p_i^k - q^w(w - c_{ij})]^2 = \sigma_w^2 \tag{4.103}$$

where the index n' is understood, and the $\delta_{n'1}$ term in Equation (4.103) indicates that housing costs are financed by heads of households. The maximum-entropy model equations will be stated for $n' = 1$ only (and with it understood, not shown explicitly). The $n' = 0$ equations will be identical, but with the final exponential term omitted. The model can be obtained by maximising

$$S = -\sum_i \sum_j \sum_k \sum_w \sum_n T_{ij}^{kwn}! \tag{4.104}$$

subject to Equations (4.97)-(4.103). The result, for $n' = 1$, can be expressed in groups according to values of n:

(a) $n = 1$: $$T_{ij}^{kw1} = A_i^{k\,1/3} B_j^{w1} r^n \Delta H_i^k \Delta E_j^{w1} \exp(-\beta^w c_{ij}) \times \exp\{-\mu^w [p_i^k - q^w(w - c'_{ij})]^2\} \; ; \tag{4.105}$$

(b) $n = 2$: $$T_{ij}^{kw2} = A_i^{k\,2/4} B_j^{w2} r^n H_i^k \Delta E_j^{w2} \exp(-\beta^w c_{ij}) \; ; \tag{4.106}$$

(c) $n = 3$: $$T_{ij}^{kw3} = A_i^{k\,1/3} B_j^{w} r^n \Delta H_i^k E_j^{w} \exp(-\beta^w c_{ij}) \times \exp\{-\mu^w [p_i^k - q^w(w - c'_{ij})]^2\} \; ; \tag{4.107}$$

(d) $n = 4$: $$T_{ij}^{kw4} = A_i^{k\,2/4} B_j^{w} H_i^k E_j^{w} \exp(-\beta^w c_{ij}) \; ; \tag{4.108}$$

where

$$A_i^{k\,1/3} = \frac{\exp(-\lambda_i^{(1/3)k})}{r^n \Delta H_i^k} \tag{4.109}$$

$$A_i^{k\,2/4} = \frac{\exp(-\lambda_i^{(2/4)k})}{r^n H_i^k} \tag{4.110}$$

$$B_j^{w1} = \frac{\exp(-\lambda_j^{(1)w})}{\Delta E_j^{w1}} \tag{4.111}$$

$$B_j^{w2} = \frac{\exp(-\lambda_j^{(2)w})}{\Delta E_j^{w2}} \tag{4.112}$$

$$B_j^{w} = \frac{\exp(-\lambda_j^{w})}{E_j^{w}} \, , \tag{4.113}$$

where $\lambda_i^{(1/3)k}$, $\lambda_i^{(2/4)k}$, $\lambda_j^{(1)w}$, $\lambda_j^{(2)w}$, λ_j^w, β^w, and μ^w are the Lagrangian multipliers associated with Equations (4.97)-(4.103) respectively. $A_i^{k\,1/3}$, $A_i^{k\,2/4}$, B_j^{w1}, B_j^{w2}, and B_j^w can be found by substitution in Equations (4.97)-(4.101) respectively.

Thus this example illustrates how the model builder can incorporate the set of improvements to the elementary models of Section 4.4.1 which he considers feasible, especially taking account of data availability.

5

Problems of missing or additional information in flow and location models

5.1 Introduction

In this chapter, we examine a set of problems which appear to have little in common at first sight, but which appear as similar problems when the related models are studied using the entropy-maximising methodology. Three examples will be considered, though many more similar problems can be tackled with identical methods. Firstly, we consider the problem of modelling transport flows for a study area which has strong interactions with a zone or zones outside it. Secondly, we consider the problem of modelling transport flows for a study area, within which there are car-parking capacity restraints imposed in some zones. These examples are concerned with the flows themselves. Thirdly, we consider the problem of a residential-location model for a study area where some zonal populations are fixed exogenously by planning policy. The similarities of approach to these different problems will become clear below.

5.2 External zones in transport models: the gravity model with missing information

In the transport-flow models outlined so far in Chapters 2 and 3, the study area has been considered to be closed; that is, it has been assumed that there is no interaction between zones in the study area and zones outside it. This is obviously a serious restriction: in many cases it is possible to draw the study area boundary in such a way that the approximation is a reasonable one, but not always. One example where the approximation is poor arises as follows: suppose there is a large town outside a study area consisting of a set of smaller towns. That is, the main purpose of the study is to analyse the interaction between the small towns, but it is important to recognise their interaction with the large town *without* having to model in detail all the interactions within the large town itself, which would enormously complicate, and make more expensive, this kind of study. This 'closure problem' is well known, of course. Various solutions have been proposed which need not concern us here. We only want to show what the maximum-entropy method contributes to the problem.

The essence of the maximum probability method of constructing gravity models is that *all* the information we have about a problem must be represented and formalised in the constraints. To fix ideas, consider our usual work-trip example. For one trip purpose, say the journey to work, we have, as in the situations of Chapter 2, the following information:

(1) numbers of workers, by person type by zone, O_i^n, and, if necessary, for external zones;

(2) number of jobs by zone, D_j, for both the study area and external zones;

(3) detailed network information, c_{ij}^k, the generalised cost of travelling from i to j by mode k for any pair of zones i, j, of which at least one is in the study area.

We do *not* have detailed c_{ij}^k information for external–external trips. It is particularly important to note this if external zones are numerous. This means that the model should not attempt to estimate trip interchanges for external to external pairs of zones. Further, external zones are usually rather large, and so trip-interchange data are only available on a coarse basis.

What is needed is a model which makes the best use of this information. Put simply, we want to model all work trips that have either (or both) residence and job ends in the study area, and not the rest. This would, of course, be a useful extension of the usual urban methods, as an alternative to the dummy-zone method, or the method of using a separate external model. In the case of a strong interaction between study area and external zones, however, it could be vital.

All this can be put more formally as follows:
let S be the study area, or region of interest;
let S' be S *plus* additional zones *to* which people in S can commute;
let S'' be S *plus* additional zones *from* which people who *work* in S can commute.

S' would usually be the same as S'', but they are separated in case the distinction is needed. In a particular case, for example, it could arise that there was no out-commuting, only in-commuting, and then S'' would simply become S, which would simplify the calculation.

We must now define constraint equations so that the usual spatial interaction variable for a transport model, T_{ij}^{kn} (the number of trips between zone i and zone j, by mode k, by persons of type n) add up to O_i^n and D_j, which are known, for i and j when both are within S. The 'cost' constraint can be applied to all $(i–j)$ trips where at least one of i or j is in S. This information can be represented formally as (in the notation of Chapter 2)

$$\sum_{j \in S'} \sum_{k \in M(n)} T_{ij}^{kn} = O_i^n, \qquad \text{for } i \in S, \tag{5.1}$$

$$\sum_{i \in S''} \sum_n \sum_{k \in M(n)} T_{ij}^{kn} = D_j, \qquad \text{for } j \in S, \tag{5.2}$$

$$\sum_{\substack{i \in S'' \\ j \in S'}\ \text{where at least one of } i, j \in S} \ \sum_{k \in M(n)} T_{ij}^{kn} c_{ij}^k = C^n, \tag{5.3}$$

for each person of type n.

There is a cost constraint for each person type, as usual, as this leads to the possibility of a different exponent for each person type. The entropy maximising model can be obtained by forming a Lagrangian $\mathcal{L}$, where

$$\mathcal{L} = -\sum_{\substack{i \in S'' \\ j \in S'}\Big\}\substack{\text{at least} \\ \text{one in } S}} \sum_{n} \sum_{k \in M(n)} \ln T_{ij}^{kn}!$$

$$-\sum_{i \in S} \sum_{n} \lambda_i^{(1)n} \Big(\sum_{j \in S'} \sum_{k \in M(n)} T_{ij}^{kn} - O_i^n \Big)$$

$$-\sum_{j \in S} \lambda_j^{(2)} \Big(\sum_{i \in S''} \sum_{n} \sum_{k \in M(n)} T_{ij}^{kn} - D_j \Big)$$

$$-\sum_{n} \beta^n \Bigg(\sum_{\substack{i \in S'' \\ j \in S'}\Big\}\substack{\text{at least} \\ \text{one in } S}} \sum_{k \in M(n)} T_{ij}^{kn} c_{ij}^k - C^n \Bigg) \tag{5.4}$$

where $\lambda_i^{(1)n}$, $\lambda_j^{(2)}$, and β^n are Lagrangian multipliers. We should now solve

$$\frac{\partial \mathcal{L}}{\partial T_{ij}^{kn}} = 0 \tag{5.5}$$

for $i \in S''$, $j \in S'$, with at least one $(i, j) \in S$. The easiest way to substitute from Equation (5.4) into Equation (5.5), taking account of the various partial summations in Equation (5.4), is to deal with three cases separately:

(a) $i \in S$, $j \in S$;
(b) $i \in S'' \notin S$ (that is, i in S'' but not in S), $j \in S$;
(c) $i \in S$, $j \in S' \notin S$.

Substitute from Equation (5.4) into Equation (5.5), for each of these three cases in turn:

(a) $i \in S$, $j \in S$: $$-\ln T_{ij}^{kn} - \lambda_i^{(1)n} - \lambda_j^{(2)} - \beta^n c_{ij}^k = 0\,, \tag{5.6}$$

giving

$$T_{ij}^{kn} = \exp(-\lambda_i^{(1)n} - \lambda_j^{(2)} - \beta^n c_{ij}^k)\,. \tag{5.7}$$

(b) $i \in S'' \notin S$, $j \in S$: $$T_{ij}^{kn} = \exp(-\lambda_j^{(2)} - \beta^n c_{ij}^k)\,. \tag{5.8}$$

Note that one of the multipliers, $\lambda_i^{(1)n}$, does not appear in this equation. It is not, in fact, defined for $i \in S'' \notin S$.

(c) $i \in S$, $j \in S' \notin S$: $$T_{ij}^{kn} = \exp(-\lambda_i^{(1)n} - \beta^n c_{ij}^k)\,, \tag{5.9}$$

where in this case the other Lagrangian multiplier $\lambda_j^{(2)}$ does not appear. $\lambda_j^{(2)}$ is not defined for $j \in S' \notin S$. We shall see that what this means is that, in each of Equations (5.8) and (5.9), one of the 'competition' terms, or balancing factors, has been dropped.

Thus, in order to introduce an attractiveness term for the origin zone i in Equation (5.8) and the destination zone j in Equation (5.9), we redefine the corresponding c_{ij}^k, using the device introduced in Section 3.3 of Chapter 3, that is by introducing 'benefits' u_i and v_j as appropriate.

Thus, for $i \in S'' \notin S$, $j \in S$,

$$c_{ij}^k \rightarrow c_{ij}^k - u_i \;, \tag{5.10}$$

and for $i \in S$, $j \in S' \notin S$,

$$c_{ij}^k \rightarrow c_{ij}^k - v_j \;. \tag{5.11}$$

This introduces factors $\exp(\beta^n u_i)$, $\exp(\beta^n v_j)$ in Equations (5.8) and (5.9), respectively. These terms can now be identified with a productiveness factor P_i^n and an attractiveness factor Q_j^n, respectively. They *could* be n-dependent as indicated. (Thus we are employing a different notation from that of Chapter 3 to make the notation more convenient for this case.) Further, for $i \in S$, $j \in S$, we can define

$$\exp(-\lambda_i^{(1)n}) = A_i^n O_i^n \tag{5.12}$$

$$\exp(-\lambda_j^{(2)}) = B_j D_j \tag{5.13}$$

in the usual way, thus introducing A_i^n and B_j as variables to replace $\exp(-\lambda_i^{(1)n})$ and $\exp(-\lambda_j^{(2)})$. and Equations (5.6), (5.8), and (5.9) can be written

$$T_{ij}^{kn} = A_i^n B_j O_i^n D_j \exp(-\beta^n c_{ij}^k) \;, \qquad i \in S,\; j \in S \tag{5.14}$$

$$T_{ij}^{kn} = P_i^n B_j D_j \exp(-\beta^n c_{ij}^k) \;, \qquad i \in S'' \notin S,\; j \in S \tag{5.15}$$

$$T_{ij}^{kn} = A_i^n O_i^n Q_j^n \exp(-\beta^n c_{ij}^k) \;, \qquad i \in S,\; j \in S' \notin S \;. \tag{5.16}$$

Values of A_i^n and B_j can be obtained in the usual way by substituting from Equations (5.14) and (5.16) in the constraint Equations (5.1) and (5.2). (The third constraint equation could be solved for the β^n's, but, as usual, we would more often obtain the β's from a calibration process.) The substitution has to be carried out carefully to ensure that the T_{ij}^{kn} appropriate to particular ranges of i, j in Equations (5.1) and (5.2) is used. We have

$$A_i^n O_i^n \left[\sum_{j \in S} B_j D_j \exp(-\beta^n C_{ij}^n) + \sum_{j \in S' \notin S} Q_j^n \exp(-\beta^n C_{ij}^n) \right] = O_i^n \tag{5.17}$$

$$B_j D_j \left[\sum_{i \in S} \sum_n A_i^n O_i^n \exp(-\beta^n C_{ij}^n) + \sum_{i \in S'' \notin S} P_i^n \exp(-\beta^n C_{ij}^n) \right] = D_j \;, \tag{5.18}$$

where, in each equation, we have defined and used

$$\exp(-\beta^n C_{ij}^n) = \sum_{k \in M(n)} \exp(-\beta^n c_{ij}^k) \;, \tag{5.19}$$

as in Chapter 2 [cf. Equation (2.56)]. Thus

$$A_i^n = \left[\sum_{j \in S} B_j D_j \exp(-\beta^n C_{ij}^n) + \sum_{j \in S' \notin S} Q_j^n \exp(-\beta^n C_{ij}^n) \right]^{-1} \tag{5.20}$$

$$B_j = \left[\sum_{i \in S} \sum_n A_i^n O_i^n \exp(-\beta^n C_{ij}^n) + \sum_{i \in S'' \notin S} \sum_n P_i^n \exp(-\beta^n C_{ij}^n) \right]^{-1} . \tag{5.21}$$

It seems intuitively clear that measures of productiveness and attractiveness for zones outside S should be, as one would expect, numbers of workers and jobs. The various terms on the right-hand side of Equations (5.20) and (5.21) would then be of the same type. However, this could conceivably be amended slightly and the P_i^n's and Q_j^n's, or at least the largest of them, could be treated as parameters of the model. These and similar adjustments could be made to improve the fit of the model to any available trip data T_{ij}^{kn}, or aggregates of this, for $i \in S''$, $j \in S'$ with at least one of i and j in S, though changes for the purpose of better fits should not be made without some reason[(12)].

Note that, for computational purposes, the P_i^n's and Q_j's can be interpreted as trip ends, and options included in the computer programs to set A_i or B_j as 1 as appropriate if i or j are external zones. It can then be seen that the model given by Equations (5.14)–(5.16) and (5.20) and (5.21) can be computed as the ordinary model of Chapter 2, with the minor adjustments noted. This facilitates the computational problems associated with handling external zones.

5.3 Parking-policy constraints in transport models

It is often stated that parking control is one of the management tools available to transport planners. However, models, as at present formulated, cannot easily incorporate this concept. It turns out that the maximum-entropy methodology lends itself to the solution of this problem.

For simplicity initially, suppose we have a closed study area (i.e. we neglect all external trips) in order to simplify the algebra. Such complications can easily be reintroduced later using the methods of Section 5.2 above. As an example, we can take the generalised distribution model of Section 2.3, obtained by maximising

$$S = \frac{\ln T!}{\prod\limits_{ijnk \in M(n)} T_{ij}^{kn}!} \tag{5.22}$$

subject to the constraints

$$\sum_j \sum_{k \in M(n)} T_{ij}^{kn} = O_i^n \tag{5.23}$$

$$\sum_i \sum_n \sum_{k \in M(n)} T_{ij}^{kn} = D_j \tag{5.24}$$

$$\sum_i \sum_j \sum_{k \in M(n)} T_{ij}^{kn} c_{ij}^k = C^n \tag{5.25}$$

(12) We might also assume, by analogy with the definition of D_j, that Q_j^n can be written as Q_j, and assumed to be independent of n.

in the usual way, and this gives

$$T_{ij}^{kn} = A_i^n O_i^n B_j D_j \exp(-\beta^n c_{ij}^k) \tag{5.26}$$

with the usual expressions for A_i^n and B_j [actually given in Equations (2.38) and (2.39); the other equations have been repeated for convenience].

To operate a parking policy would be to fix what we can define as D_j^k—the number of trip attractions by a mode k (in this case car) in zone j. It will be useful to leave the notation in general form in case similar constraints arise with other modes (due, for example, to rail-station capacities). Then let P be the set of (i, k) pairs for which some constraint exists. That is, we add to the constraint Equations (5.23)–(5.25) the constraint

$$\sum_i \sum_n T_{ij}^{kn} = D_j^k \text{ for } (j, k) \in P \ . \tag{5.27}$$

Thus, we have to maximise a Lagrangian $\mathcal{L}$, where

$$\begin{aligned}\mathcal{L} = & -\sum_i \sum_j \sum_n \sum_{k \in M(n)} \ln T_{ij}^{kn}! + \sum_j \lambda_i^{n(1)} \Big(O_i^n - \sum_j \sum_{k \in M(n)} T_{ij}^{kn} \Big) \\ & + \sum_j \lambda_j^{(2)} \Big(D_j - \sum_j \sum_n \sum_{k \in M(n)} T_{ij}^{kn} \Big) + \sum_{jk \in P} \mu_j^k \Big(D_j^k - \sum_i \sum_{n \in m(k)} T_{ij}^{kn} \Big) \\ & + \sum_n \beta^n \Big(C^n - \sum_i \sum_j \sum_{k \in M(n)} T_{ij}^{kn} c_{ij}^k \Big) \ . \end{aligned} \tag{5.28}$$

Note that, in the fourth term on the right-hand side of Equation (5.27), a new set of Lagrangian multipliers μ_j^k has been introduced, associated with Equation (5.27), for $(i, k) \in P$, and note that the n summation in the same term is over a set $m(k)$, which is the set of n's which have mode k available to them [the 'inverse' of $M(n)$]. This maximum can be obtained by solving

$$\frac{\partial \mathcal{L}}{\partial T_{ij}^{kn}} = 0 \ , \qquad (j, k) \notin P \tag{5.29}$$

$$\frac{\partial \mathcal{L}}{\partial T_{ij}^{kn}} = 0 \ , \qquad (j, k) \in P \tag{5.30}$$

and this gives

$$T_{ij}^{kn} = \exp(-\lambda_i^{n(1)} - \lambda_j^{(2)} - \beta^n c_{ij}^k) \ , \qquad (j, k) \notin P \tag{5.31}$$

$$T_{ij}^{kn} = \exp(-\lambda_i^{n(1)} - \lambda_j^{(2)} - \mu_j^k - \beta^n c_{ij}^k) \ , \qquad (j, k) \in P \ . \tag{5.32}$$

Put

$$\exp(-\lambda_i^{n(1)}) = A_i^n O_i^n \tag{5.33}$$

$$\exp(-\lambda_j^{(2)}) = B_j D_j \tag{5.34}$$

in the usual way, and substitute from Equations (5.31) and (5.32) into Equations (5.23) and (5.24), respectively, and using Equations (5.33) and (5.34) gives

$$A_i^n = \left[\sum_j \sum_{\substack{k \in M(n) \\ \notin P}} B_j D_j \exp(-\beta^n c_{ij}^k) + \sum_j \sum_{\substack{k \in M(n) \\ \in P}} B_j D_j \exp(-\mu_j^k - \beta^n c_{ij}^k)\right]^{-1} \tag{5.35}$$

$$B_j = \left[\sum_i \sum_n \sum_{k \in M(n)} A_i^n O_i^n \exp(-\beta^n c_{ij}^k)\right]^{-1}, \qquad (i, k) \notin P \tag{5.36}$$

$$B_j = \left[\sum_i \sum_n \sum_{k \in M(n)} A_i^n O_i^n \exp(-\mu_j^k - \beta^n c_{ij}^k)\right]^{-1}, \qquad (j, k) \in P\,. \tag{5.37}$$

μ_j^k can be obtained by substituting from Equations (5.31) and (5.32) into Equation (5.27) to give

$$\exp(-\mu_j^k) = D_j^k\left[\sum_i \sum_{n \in m(k)} A_i^n O_i^n B_j D_j \exp(-\beta^n c_{ij}^k)\right]^{-1}. \tag{5.38}$$

It turns out to be convenient to leave μ_j^k in this form rather than to use transformations of the form of Equations (5.33) and (5.34), as it is found to have more of the nature of a cost than a 'balancing factor'. (Note that, of course, $\lambda_i^{n(1)}/\beta^n$ and $\lambda_j^{(2)}/\beta^n$ can be interpreted as costs, but that is by the way.)

In principle, the whole model could be solved iteratively. 'Guess' starting values for μ_j^k and B_j, calculate A_i^n, calculate B_j, calculate μ_j^k, and repeat. However, in computational practice, this would mean writing a new and rather complicated computer program. An alternative is to make an approximation which also helps in the interpretation of μ_j^k.

Note that, in Equation (5.38), if μ_j^k is zero, then, formally, the equation gives

$$D_j^k = \sum_i \sum_{n \in m(k)} A_i^n O_i^n B_j D_j \exp(-\beta^n c_{ij}^k)\,, \tag{5.39}$$

which is just the number of trip attractions in j by mode k which the non-parking-constrained model gives. This does not mean the situation is necessarily entirely unconstrained, but that the only constraint is the D_j constraint in Equation (5.24); the constraint Equation (5.27) is redundant. As μ_j^k increases from zero, the constrained D_j^k is seen to be less than that which would be given in the non-parking-constrained model. This suggests that μ_j^k, or to be more precise μ_j^k/β^n, can be interpreted as the 'cost' of imposing the parking constraint D_j^k. This means also, and perhaps more usefully, that it is the average additional parking cost which should be imposed to ensure that the D_j^k constraint is satisfied. In this sense, the concept has an obvious operational use for transportation planners.

In an actual application, note that the = sign in Equation (5.27) should be, strictly, $\leqslant$. However, the mathematical problems associated with handling inequalities of this kind are very great (cf. Kuhn and Tucker, 1956). In practice, constraints would only be applied in central city zones where it is known that, without constraints, the D_j^k the planner has in mind would be exceeded. So, in practice, we can assume that the equality sign holds for the zones where a constraint is being applied.

Suppose now that there are relatively few such zones. Then it could be argued that the A_i^n's and B_j's in Equations (5.35) and (5.37), *with* μ_j^k *set equal to zero*, are near enough correct. These are, in fact, the usual A_i^n's and B_j's. Equation (5.38) could then be solved to give an estimate for the μ_j^k's, using the approximate A_i^n's and B_j's. Thus the usual computer programs can be used, though a modification might be necessary to output explicitly the A_i^n's and B_j's, and one auxiliary program would be needed to calculate μ_j^k using Equation (5.38). The quantity μ_j^k/β^n could then be added to c_{ij}^k, and the next run of the model would give, approximately, the parking-constrained distribution of trips and associated modal splits. Thus such a calculation, involving little computation other than what would be carried out anyway, should give a reasonable description for the planner of the consequences of imposing some parking constraints.

Note, incidentally, in comparing Equations (5.26) and (5.38) that Equation (5.38) can be written in the form

$$\exp(-\mu_j^k) = \frac{D_j^k}{T_{*j}^{k*}} , \tag{5.40}$$

where T_{*j}^{k*} is the total number of trips to j by mode k when there are no parking constraints. Thus, an *approximate* ('approximate' because the A_i^n's and B_j's used to estimate T_{*j}^{k*} assume $\mu_j^k = 0$) value of the cost to be imposed to achieve constrained totals D_j^k in j is

$$\frac{\mu_j^k}{\beta^n} = \frac{1}{\beta^n} \ln \frac{T_{*j}^{k*}}{D_j^k} , \tag{5.41}$$

which could provide the basis for useful back-of-envelope calculations.

5.4 Policy and capacity constraints in residential location

The third problem was that of coping with zones whose populations had been exogenously fixed (say by planners). Let Z_1 be the set of zones i whose populations are exogenously fixed, and Z_2 be the set of zones i whose populations are still to be determined by the model. Then the union of these sets Z, given by

$$Z = Z_1 \cup Z_2 , \tag{5.42}$$

is the set of all zones. The fact that P_i is fixed for $i \in Z_1$ can be represented by the constraint equation

$$\sum_{j \in Z} T_{ij} = P_i , \qquad i \in Z_1 , \tag{5.43}$$

and we have as usual [cf. Equations (4.17) and (4.14)]

$$\sum_{i \in Z} T_{ij} = E_j \,, \qquad j \in Z \tag{5.44}$$

$$\sum_{i \in Z} \sum_{j \in Z} T_{ij} c_{ij} = C \,. \tag{5.45}$$

It is possible to maximise $\ln T! \big/ \prod_{ij} T_{ij}!$ to get a Lowry kind of residential-location model, modified by the addition of the constraint Equation (5.43) (cf. Section 4.4.2 in the preceding chapter).

We can now maximise a Lagrangian $\mathcal{L}$, where

$$\mathcal{L} = \ln T! - \sum_{ij} \ln T_{ij}! + \sum_{i \in Z_1} \lambda_i^{(1)} \left(P_i - \sum_{j \in Z} T_{ij} \right) + \sum_{j \in Z} \lambda_j^{(2)} \left(E_j - \sum_{i \in Z} T_{ij} \right) + \beta \left(C - \sum_{i \in Z} \sum_{j \in Z} T_{ij} c_{ij} \right) \,, \tag{5.46}$$

and where $\lambda_i^{(1)}$, $\lambda_j^{(2)}$, and β are Lagrangian multipliers associated with Equations (5.42), (5.43), and (5.44), respectively. Note that the summations have to be defined explicitly and carefully. We have to solve

$$\frac{\partial \mathcal{L}}{\partial T_{ij}} = 0 \tag{5.47}$$

and the constraint Equations (5.42)–(5.44). Equation (5.47) gives

$$-\ln T_{ij} - \lambda_i^{(1)} - \lambda_j^{(2)} - \beta c_{ij} = 0 \text{ for } i \in Z_1 \tag{5.48}$$

and

$$-\ln T_{ij} - \lambda_j^{(2)} - \beta c_{ij} = 0 \text{ for } i \in Z_2 \,. \tag{5.49}$$

So

$$T_{ij} = \exp(-\lambda_i^{(1)} - \lambda_j^{(2)} - \beta c_{ij}) \,, \qquad i \in Z_1 \tag{5.50}$$

and

$$T_{ij} = \exp(-\lambda_j^{(2)} - \beta c_{ij}) \,, \qquad i \in Z_2 \,. \tag{5.51}$$

In the usual way

$$\exp(-\lambda_i^{(1)}) = A_i P_i \,, \qquad i \in Z_1 \tag{5.52}$$

$$\exp(-\lambda_j^{(2)}) = B_j E_j \,, \qquad j \in Z \tag{5.53}$$

and the model, as represented in Equations (5.50) and (5.51), can be written

$$T_{ij} = A_i B_j P_i E_j \exp(-\beta c_{ij}) \,, \qquad i \in Z_1 \tag{5.54}$$

$$T_{ij} = B_j E_j \exp(-\beta c_{ij}) \,, \qquad i \in Z_2 \,, \tag{5.55}$$

where

$$A_i = \left[\sum_{j \in Z} B_j E_j \exp(-\beta c_{ij}) \right]^{-1}, \qquad i \in Z_1 \tag{5.56}$$

$$B_j = \left[\sum_{i \in Z_1} A_i P_i \exp(-\beta c_{ij}) + \sum_{i \in Z_2} \exp(-\beta c_{ij}) \right]^{-1}, \qquad j \in Z. \tag{5.57}$$

Note that an A_i has been introduced for each zone with an exogenously determined population, and there is one B_j for each zone as usual.

This development can be combined with that of Chapter 3, and applied to the residential model in Section 4.4.2 giving Equations (4.21) and (4.22): V_i^α's can be introduced to measure and represent the inherent relative attractiveness of i, so

$$T_{ij} = A_i B_j P_i V_i^\alpha E_j \exp(-\beta c_{ij}), \qquad i \in Z_1 \tag{5.58}$$

$$T_{ij} = V_i^\alpha B_j E_j \exp(-\beta c_{ij}), \qquad i \in Z_2, \tag{5.59}$$

where

$$A_i = \left[\sum_j B_j E_j \exp(-\beta c_{ij}) \right]^{-1}, \qquad i \in Z_1 \tag{5.60}$$

$$B_j = \left[\sum_{i \in Z_1} A_i P_i V_i^\alpha \exp(-\beta c_{ij}) + \sum_{i \in Z_2} V_i^\alpha \exp(-\beta c_{ij}) \right]^{-1}, \qquad j \in Z. \tag{5.61}$$

If, however, V_i^α is taken as the existing size of the zone, then V_i^α can be taken to be 1 for $i \in Z_1$, as the effect is already included, and as P_i (or a similar measure) for $i \in Z_2$. These equations would demand iterative solution unless the P_i's used for V_i^α were those of a previous time period.

A related problem, as in the case of parking constraints, is that of dealing with capacity constraints. This is the problem of recognising zonal population capacities. It is not identical because the constraints are essentially inequalities and not equalities.. Let $P_i^{(c)}$ be the population capacity of i, and the equivalent of Equation (5.43) is

$$\sum_j T_{ij} \leqslant P_i^{(c)}. \tag{5.62}$$

The usual method of dealing with capacity constraints in elementary location models is to check $\sum_j T_{ij}$ against $P_i^{(c)}$, and if the inequality Equation (5.62) is infringed to reallocate the excess among other zones in accordance with some rule (either in proportion to population or in proportion to growth). If constraints are infringed a second time (in other zones), the process is iterated until all the capacity constraints are satisfied. The difficulty of this method is that the final solution is not an equilibrium one, in the sense that the final T_{ij} does not satisfy the basic equations such as Equations (4.21) and (4.22). It is difficult to see what the behavioral implications are of such a procedure. If the constraints are not very demanding, and are only applicable to a few zones, then these criticisms probably matter only a little; if not, other methods should be used.

One possibility is some kind of linear programming model [whose dual will also produce 'location' rents determined by the capacity constraint equations (Stevens, 1961)]. There is the usual difficulty, however, that it is difficult to specify the objective function sufficiently well.

A second possibility is to use the results just obtained, which dealt with the case where zone populations are fixed exogenously. Suppose after an iteration the constraint Equation (5.62) is infringed for a number of zones. Then we *know* that this zone would reach capacity and these zones could be switched from the set Z_2 to Z_1 for the next iteration. All zones would be in Z_2 initially. This procedure has now been used in a model by Batty (1970). Using this procedure, the final solution will

(1) not infringe any capacity constraints;
(2) not overestimate the attractiveness of zones which do not ultimately reach capacity (as attractiveness is proportional to size);
(3) satisfy a set of equilibrium equations.

6

The use of entropy in the analysis of utility-maximising systems

6.1 Introduction
This chapter is something of a digression. So far in this book, we have been largely concerned with spatial interaction variables, such as T_{ij}, the number of work trips from zone i to zone j. The purpose of this chapter is to investigate the application of entropy in a different kind of system—a utility-maximising system. This is of interest in itself, but will also enable us to connect to some of the general notions of Chapter 1, and finally to investigate the status of systems such as transport systems (and, generally, urban and regional systems) as utility-maximising systems; this has relevance to evaluation issues.

6.2 Utility-maximising systems
A good example of a utility-maximising system is well known in the theory of consumers' behaviour (see, for example, Henderson and Quandt, 1958; Samuelson, 1947, etc.). The usual statement, for a single consumer, runs something like the following: let $x_1, x_2, ..., x_N$ [13] be the quantities of goods 1, 2, ..., N purchased by the consumer at prices $p_1, p_2, ..., p_N$ out of an income I. The consumer maximises his utility

$$u = u(x_1, x_2, x_3, ..., x_N, I) \tag{6.1}$$

subject to his budget constraint

$$\sum_i x_i p_i = I \; . \tag{6.2}$$

Now, define a Lagrangian $\mathcal{L}$

$$\mathcal{L} = u(x_1, x_2, ..., x_N, I) + \lambda\left(I - \sum_i p_i x_i\right), \tag{6.3}$$

where λ is the multiplier associated with Equation (6.2). So, maximising in the usual way, the amount of the good i which is purchased is the solution of

$$\frac{\partial \mathcal{L}}{\partial x_i} = \lambda p_i \; . \tag{6.4}$$

This solution can be written in the form

$$x_i = x_i(p_1, p_2, ..., p_N, I) \tag{6.5}$$

and is known as the demand function for i.

(13) Note that x_N can always be taken as 'money' and $x_1, x_2, ..., x_{N-1}$ the goods of particular interest in the current analysis.

It can also be shown that, at a given utility level $\bar{u}$,

$$\left.\frac{\partial I}{\partial p_i}\right|_{u=\bar{u}} = x_i \ , \tag{6.6}$$

and this equation is used to evaluate the consequences of price changes on the consumer's income (cf. Wilson and Kirwan, 1969).

The calculations can be repeated for a whole set of consumers. A single consumer can be labelled by k and Equations (6.1), (6.2), (6.4), (6.5), and (6.6) can be written

$$u^{(k)} = u^{(k)}(x_1^{(k)}, x_2^{(k)}, ..., x_N^{(k)}, I^{(k)}) \tag{6.7}$$

$$\sum_i x_i^{(k)} p_i = I^{(k)} \tag{6.8}$$

$$\frac{\partial u^{(k)}}{\partial x_i^{(k)}} = \lambda^{(k)} p_i \tag{6.9}$$

$$x_i^{(k)} = x_i^{(k)}(p_1, p_2, ..., p_N, I^{(k)}) \tag{6.10}$$

and

$$\left.\frac{\partial I^{(k)}}{\partial p_i}\right|_{u^{(k)} = \bar{u}^{(k)}} = x_i^{(k)} \ , \tag{6.11}$$

using an obvious notation. We can then define

$$I = \sum_k I^{(k)} \tag{6.12}$$

as the society's income, and

$$x_i = \sum_k x_i^{(k)} \tag{6.13}$$

as the aggregate demand for the good i, and, under suitable conditions, Equation (6.11) can be summed over k to give

$$\frac{\partial I}{\partial p_i} = x_i \ . \tag{6.14}$$

Equation (6.14) is identical in form with Equation (6.6). Thus, in suitable circumstances, Equations (6.1)–(6.6) can be interpreted as the equations describing a single consumer in a utility-maximising system, or a group of consumers. The condition under which the transition can be made from a single consumer to a group are stringent but need not concern us here. Henceforth, then, we shall take Equations (6.1)–(6.6) to illustrate the notion of a utility-maximising system, while being conscious that this is an oversimplification in economic terms.

6.3 The use of entropy in the analysis of utility-maximising systems

In order to define the entropy of the system of interest, we need to define some suitable variables. We saw in Chapter 1 that the concept of entropy is best associated with a probability distribution. It is instructive, as a preliminary, to review more fundamentally how we constructed the entropy of systems in previous chapters. Consider a transport system where T_{ij} is the number of trips from zone i to zone j. The total number of trips

$$T = \sum_i \sum_j T_{ij} \tag{6.15}$$

has always been taken as fixed and given, and we shall see later that this is of some significance. The entropy can be taken in any of the following forms:

$$\ln \frac{T!}{\prod_i \prod_j T_{ij}!} \tag{6.16}$$

$$-\sum_i \sum_j T_{ij} \ln T_{ij} \tag{6.17}$$

or

$$-\sum_i \sum_j p_{ij} \ln p_{ij} \ , \tag{6.18}$$

where

$$p_{ij} = \frac{T_{ij}}{T} \ , \tag{6.19}$$

and p_{ij} can be interpreted as a probability distribution. In this case, it has been shown that all the measures of entropy are equivalent.

We can now reconsider the utility-maximising system set out in Section 6.2 above. The first problem is that the quantities $x_1, x_2, ..., x_N$ may be measured in different units. Thus, the obvious analogues of the expressions (6.16)–(6.18) could not be interpreted as entropy. Even if the units of measurement were identical, and we defined

$$x = \sum_i x_i \tag{6.20}$$

and

$$y_i = \frac{x_i}{x} \ , \tag{6.21}$$

it is not clear that

$$S = -\sum_i y_i \ln y_i \tag{6.22}$$

is a suitable measure of entropy because x is not fixed—we have to find a way of allowing x to be a variable.

The appropriate fixed quantity in this case is I, and we can define

$$y_i = \frac{x_i p_i}{I} \tag{6.23}$$

as the proportion of consumer income spent on the good i. Then Equation (6.22) serves as an appropriate definition of entropy. With this definition, it does not matter that the x_i's are measured in different units. The variables y_i thus defined have some of the characteristics of market shares (cf. Blackburn, 1969).

The utility-maximising system can now be characterised in terms of y_i as follows:

$$u = u\left(\frac{y_1 I}{p_1}, \frac{y_2 I}{p_2}, \dots, \frac{y_N I}{p_N}, I\right) \tag{6.24}$$

$$\sum_i y_i = 1 \tag{6.25}$$

$$\mathcal{L} = u\left(\frac{y_1 I}{p_1}, \dots\right) + \lambda\left(1 - \sum_i y_i\right), \tag{6.26}$$

so

$$\frac{\partial u}{\partial y_i} = \lambda \tag{6.27}$$

$$y_i = y_i(p_1, p_2, \dots, p_N, I), \tag{6.28}$$

and

$$\left.\frac{\partial I}{\partial p_i}\right|_{u=\bar{u}} = \frac{y_i I}{p_i}. \tag{6.29}$$

It can easily be checked [for example, by choosing some specific u and making the transformation from x_i to y_i using Equation (6.23)] that Equations (6.24)–(6.29) describe the same system as Equations (6.1)–(6.6). [It can also be checked that the transformation Equation (6.21) with the addition of Equation (6.20) to the budget constraint as an additional constraint also gives an equivalent description, but one which is very much more clumsy algebraically].

Suppose we now use the entropy-maximising approach to analyse this system. We would identify a number of constraints of the form

$$f_k(y_1, y_2, \dots, y_N) = g_k, \tag{6.30}$$

where, for convenience, all terms including y_i's are included in f_k, and others, 'constants', in g_k, and maximise the entropy S given by Equation (6.22) subject to Equation (6.25) (in this case the budget constraint) and the constraint Equation (6.30). That is, we form a Lagrangian

$$\mathcal{L} = S + \lambda\left(1 - \sum_i y_i\right) + \sum_k \mu_k (g_k - f_k) \tag{6.31}$$

and solve

$$\frac{\partial \mathcal{L}}{\partial y_i} = 0, \tag{6.32}$$

together with Equations (6.25) and (6.30). This gives

$$-\ln y_i - 1 - \lambda - \sum_k \mu_k \frac{\partial f_k}{\partial y_i} = 0 \ ,$$

and so, absorbing the 1 into the λ, we have

$$\ln y_i = -\lambda - \sum_k \mu_k \frac{\partial f_k}{\partial y_i} \ . \tag{6.33}$$

Suppose that all the y_i's, p_i's, and I have been correctly specified and that the set of constraints, Equation (6.30), have been correctly identified, so that y_i in Equation (6.33) gives a correct model for the y_i's, which is checked against observation. Then we can note that the entropy-maximising problem, in this case, is formally equivalent to maximising a quantity[(14)]

$$u = S + \sum_k \mu_k (g_k - f_k) \tag{6.34}$$

subject to the constraint Equation (6.25), and so, if the system is a utility-maximising system, we can identify u given by Equation (6.34) as the utility.

If (as is likely for most systems) the entropy S plays no role in the utility function, this will exhibit itself in one of two ways:

(1) the parameters μ_k will be very large compared with 1—large enough to reduce S in Equation (6.34) to insignificance;

or

(2) there will be as many constraints (components of the utility function) as there are variables y_i, in which case the set of constraint equations can be solved directly for the y_i's without reference to the entropy. Any such problem can always be converted to a maximisation problem if it is desired to construct a utility function (see, for example, Samuelson, 1947). (The reader can easily check that maximising entropy in such a case lengthens the algebra but gives the same answer.)

Thus, as one would expect, the entropy maximiser and the analyst of the utility-maximising system (the former postulating constraints and testing the outcomes, the latter postulating utility functions and testing the outcomes) will eventually arrive at the same answer. The entropy maximiser has in this case the same three advantages which he has over the statistician, as outlined in Chapter 1:

(a) he may have some *a priori* knowledge or intuitions about some constraints on the y_i's and this will help him to build a consistent picture of the utility function;

(14) It was pointed out to me by Shlomo Angel that it should be, and can be, checked that maximising $u(y_1, y_2, ...)$ of Equation (6.34) subject to $\sum_i y_i = 1$ is equivalent to maximising $u(x_1, x_2, ...)$ subject to $\sum_i x_i p_i = I$. This is left as an exercise to the reader.

(b) he can interpret the constraints individually (not much of an advantage in this case as the terms of the utility function can in any case be individually interpreted);

and

(c) he can be aided in the construction of dynamic models.

6.4 Urban systems as utility-maximising systems

6.4.1 Transport systems

It is a simple exercise for the reader to take any of the maximum entropy transport models presented in this book and present it as an equivalent utility-maximising system. In fact, since it is an equation of the form of Equation (6.6), which is used in evaluation procedures, the actual representation of the transport system as a utility-maximising system is superfluous—all that is needed is the demand function in the form of Equation (6.5).

Consider, for example, the transport model given by Equations (2.21)-(2.23) in Chapter 2. In the formalism of Equation (6.5), T_{ij} is known in the form

$$T_{ij} = T_{ij}(c_{11}, c_{12}, ..., \beta, E) , \tag{6.35}$$

where E stands for the other given parameters such as the O_i's and D_j's. T_{ij} is not explicitly a function of income, but β is likely to be a function of income. Then

$$\frac{\partial I}{\partial c_{ij}} = T_{ij} \tag{6.36}$$

can be used in evaluation procedures (cf. Wilson and Kirwan, 1969).

Note that each possible trip, for each (i, j) pair, is considered as a separate good. Further, implicitly, only transport goods have been considered. This question will be taken up in the next section.

6.4.2 Urban systems

In the previous section, the possibility of 'buying' a trip from i to j was regarded as the possibility of buying a single good at price c_{ij}. In the same way, we could consider the purchase of a house in zone i at price h_i (assuming, for simplicity, that all houses in zone i have the same price) as the purchase of a good, and acquiring a job in zone j offering wage w_j can be considered a third kind of good. The individual k's utility function in this paradigm could be written in the form

$$u^k = u^k(x_1, x_2, ..., y_1, y_2, ..., z_{11}, z_{12}, ...) , \tag{6.37}$$

where the x's, y's, and z's can take the value 0 or 1 (so that the maximisation problem becomes an integer programming problem) and indicate purchase of house, acquiring a job, and purchasing a work trip, respectively. A number of constraints could be written down relating the variables, including a budget constraint, and a restriction that the work trip is from home to job. (Actually, one set of variables is superfluous, but all will be retained for ease of exposition.)

An alternative way of representing this paradigm is to postulate that there are three urban goods—houses, jobs, and journey to work—and that the individual k in acquiring quantities h, w, and c has utility

$$u^k = u^k(h, w, c) , \tag{6.38}$$

where the indivisibilities enable us to use the costs as measures of the quantities bought. [A more appropriate functional form might be

$$u^k = u^k(w-h-c, h, c) , \tag{6.39}$$

but Equation (6.38) is sufficiently general to include this as a possibility.] In this case, the sets of quantities h_i, w_j, and c_{ij} can be regarded as the 'packages' within which the goods are offered. Thus, if individual k chooses to live in zone i and work in zone j, his utility is

$$u^k = u^k(h_i, w_j, c_{ij}) . \tag{6.40}$$

The models presented in this book are not primarily constructed as models of microeconomic behaviour (though they can be interpreted as such, using probabilities). We have been primarily interested in residential distribution, workplace distribution, trip distributions, and so on. In retaining this interest, the next step is to find methods of aggregating the two representations presented in this section and to explore further the interrelationships of utility analysis and entropy-maximising analysis.

Firstly, let us return to the utility function given by Equation (6.37). For each individual, we have a utility function defined in the space of the goods which have been postulated, and a budget constraint, which is a hyperplane in this space. The utility-maximisation problem is to find the indifference surface to which the hyperplane is a tangent, for each individual. To simplify our example, suppose all individuals have identical budget constraints. Assume, however, that individuals have different utility functions, which can be expressed in the following form [which is more restrictive than Equation (6.37)]:

$$u^k = u(x_1, x_2, ..., y_1, y_2, ..., z_{11}, z_{12}, ..., a_1^k, a_2^k, ...) , \tag{6.41}$$

where the a_i's are parameters which can take a different value for each individual. That is, we are assuming that the utility functions have the same functional form in the x's, y's, and z's, and that the differences arise through different values of the parameters a_i. We can solve the maximisation problem in the usual way to give (taking x_i^k as an example):

$$x_i^k = x_i^k(h_1, h_2, ..., w_1, w_2, ..., c_{11}, c_{12}, ..., a_1^k, a_2^k, ...) . \tag{6.42}$$

However, we are interested in the quantity

$$X_i = \sum_k x_i^k \ . \tag{6.43}$$

This can be calculated as follows. Assume that each a_i has a known probability distribution $f_i(a_i)$. Then, for each value of x_i^k in Equation (6.42), Equation (6.42) defines a surface, which we can call $A(x_i^k)$ in the space defined by the parameters a_i. Let $F(x_i)$ be the probability distribution of x_i (dropping the k). Then F can be calculated from the f_i's [Lindley, 1965, p.117, Equation (4)] as follows:

$$F(x_i) = \int_{A(x_i)} \cdots \int \prod_i f_i(a_i) \mathrm{d}a_i \ , \tag{6.44}$$

where the integral is taken over the surface $A(x_i)$. X_i is then simply $X_i = XF(x_i)$, where X is the total population.

It only needs a little thought to see that, even with a very simple postulated form for the function u, the surface $A(x_i)$ can be complicated and the integral of Equation (6.44) difficult to evaluate. This may go some way to explaining why it is difficult to develop urban models from first principles on a utility-maximising basis (especially as we are considering a simple paradigm anyway). Entropy-maximising models of the form presented in this book set up constraints directly on the variables such as x_i, and lead to a direct estimate of $F(x_i)$. In our paradigm, they produce density distribution of points on the budget hyperplane (defined earlier) subject to any known constraints on these distributions. This is illustrated for an even simpler two-good world in Figure 6.1.

The sets of curves $S_1, S_2, S_3, \ldots$ are the sets of indifference curves for individuals 1, 2, 3, ... AB is the budget line (the same in each case). $P_1, P_2, P_3, \ldots$ are the points where AB touches an indifference curve. What an entropy-maximising model does is to establish the density distribution of the points P_i on AB directly, subject to known constraints. In the usual absence of knowledge of the sets of indifference curves, this should be the best available procedure to obtain an approximate answer. The answer can be refined using the methods outlined in Chapter 1.

We can now turn to the second representation of the paradigm given by utility functions of the form of Equations (6.38) and (6.40). This form of utility function is not really very different from the first one. As most (all but three) of the variables in the function Equation (6.37) will be zero, one would expect the functional form of the three non-zero ones to be similar to, or identical with, that in the Equation (6.38) form. However, the form does itself suggest another kind of maximum-entropy model, and a second possible interpretation of the role of entropy-maximising analysis in relation to utility-maximising systems. This alternative arises as follows: as an example, choose a specific form of utility function, and suppose that individual k choosing to live in i and work in j achieves utility

$$u^k = a_1'(w_j - h_i - c_{ij}) + a_2' h_i + a_3' c_{ij} , \tag{6.45}$$

which can be rewritten, with redefined coefficients, as

$$u^k = a_1 h_i + a_2 w_j + a_3 c_{ij} . \tag{6.46}$$

We now face the aggregation problem. Assume that individuals, by their choices, achieve a range of utilities, but with mean $\bar{u}$. Let T be the total population, and let T_{ij} live in zone i and work in zone j in the usual way. Then we can write as a constraint for a maximum-entropy model

$$\sum_i \sum_j T_{ij}(a_1 h_i + a_2 w_j + a_3 c_{ij}) = T\bar{u} , \tag{6.47}$$

and maximising entropy in the usual way gives

$$\frac{T_{ij}}{T} = \frac{\exp[-\alpha(a_1 h_i + a_2 w_j + a_3 c_{ij})]}{\sum_i \sum_j \exp[-\alpha(a_1 h_i + a_2 w_j + a_3 c_{ij})]} \tag{6.48}$$

where α is the Lagrangian multiplier associated with Equation (6.47).

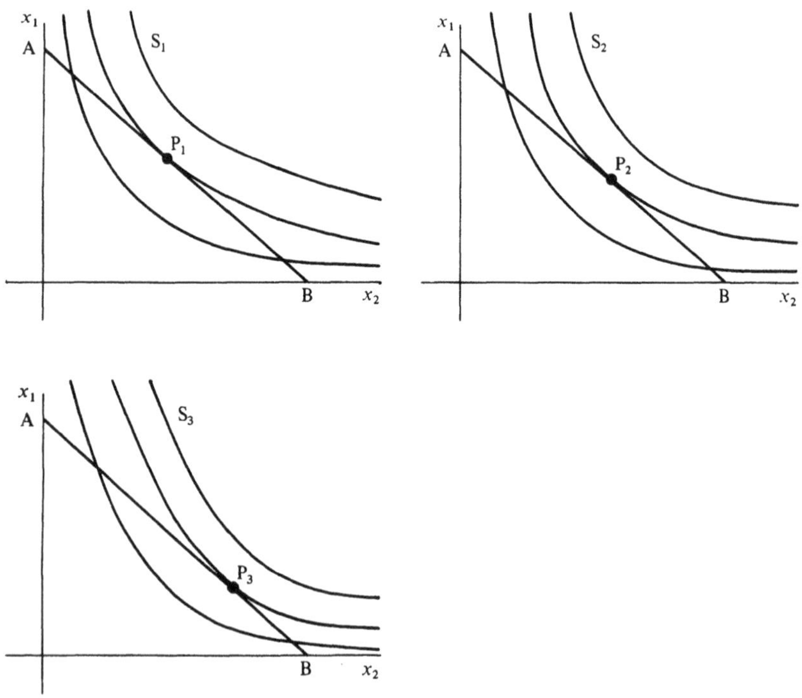

Figure 6.1.

This can be rewritten as

$$\frac{T_{ij}}{T} = \frac{\exp(-\lambda h_i - \mu w_j - \beta c_{ij})}{\sum_i \sum_j \exp(-\lambda h_i - \mu w_j - \beta c_{ij})} , \tag{6.49}$$

with suitable definitions for λ, μ, and β in terms of α, a_1, a_2, and a_3. [This form of joint residential–workplace location model was proposed in an earlier paper (Wilson, 1967b).]

It can easily be checked that the model given by Equation (6.49) would be obtained by assuming the constraints

$$\sum_i \sum_j T_{ij} h_i = H \tag{6.50}$$

$$\sum_i \sum_j T_{ij} w_j = W \tag{6.51}$$

$$\sum_i \sum_j T_{ij} c_{ij} = C , \tag{6.52}$$

when λ, μ, and β would be the Lagrangian multipliers associated with Equations (6.50)–(6.52), respectively. In normal circumstances, the coefficients a_1, a_2, and a_3 in Equation (6.47) would not be known. λ, μ, and β, on the other hand, can be determined from a knowledge of H, W, and C. They can be estimated for a base-year situation by a calibration procedure. To obtain forecasts, it is tempting to assume that a particular form of aggregation of the utility function Equation (6.40) is possible,

$$u = u(H, W, C) , \tag{6.53}$$

as a societal utility function relating total expenditure on housing H, total wages W, and total expenditure to work C.

6.4.3 Notes towards further generalisations

The bulk of Section 6.4 has been based on a paradigm which is obviously unrealistic. However, it offers insights into the analysis of utility-maximising systems in an urban and regional context, into the development of certain kinds of entropy-maximising models, and the interrelationships between the two.

This chapter concludes with some notes on how to generalise the paradigm and methods of analysis which have been outlined. The first obvious point is that all the directions of generalisation outlined for residential-location models in Chapter 4 can be applied here. For example, classification of people into income groups would be preferable to the assumption of the fixed budget constraint (which is not easily compatible with wage variation w_j!)

There is an interesting alternative disaggregation of a type not so far discussed in this book which is worth a slightly more detailed statement:

areal disaggregation in constraints of the form of Equations (6.50)–(6.52). This would seem especially reasonable in the task of modelling an urban area which is changing rather slowly. The implication is that the area in which people live or work stands as a proxy for personal characteristics such as income. Thus Equations (6.50)–(6.52) can be modified to assume that we know the amount spent on housing by residents of zone i, H_i, the wages earned by those working in zone j, W_j, and the journey-to-work expenditure by residents of i, C_i. The double summations in the constraints become single summations, and can be written

$$\sum_j T_{ij} h_i = H_i \tag{6.54}$$

$$\sum_i T_{ij} w_j = W_j \tag{6.55}$$

$$\sum_j T_{ij} c_{ij} = C_i \tag{6.56}$$

and we now see that Equations (6.54) and (6.55) can be written in the even simpler form

$$\sum_j T_{ij} = \frac{H_i}{h_i} \tag{6.57}$$

$$\sum_i T_{ij} = \frac{W_j}{w_j} \; . \tag{6.58}$$

Maximising entropy subject to the constraint Equations (6.56)–(6.58) gives, in the usual way,

$$T_{ij} = A_i B_j \frac{H_i}{h_i} \frac{W_j}{w_j} \exp(-\beta_i c_{ij}) \; , \tag{6.59}$$

where

$$A_i = \left[\sum_j \frac{W_j}{w_j} \exp(-\beta_i c_{ij}) \right]^{-1} \tag{6.60}$$

and

$$B_j = \left[\sum_i \frac{H_i}{h_i} \exp(-\beta_i c_{ij}) \right]^{-1} \; . \tag{6.61}$$

Thus we have a model of a very similar form, with the quantities H_i/h_i and W_j/w_j replacing O_i and D_j. A study of the time behaviour of the quantities H_i/h_i and W_j/w_j could give a useful lead to producing more elastic transport models (in this case for the journey to work).

Further, an addition of another constraint of the form

$$\sum_i T_{ij} c_{ij} = C'_j \tag{6.62}$$

would lead to a modified gravity model of the types suggested by Edens and Halder (see Edens, 1970; Halder, 1970; Wilson, 1970).

In conclusion, then, we note that this chapter has perhaps opened up more research questions than it has provided answers, but it is hoped that it contains interesting guidelines for further work.

7

Entropy, social physics, and general systems theory

7.1 Introduction

So far in this book we have used the concept of entropy in ways which are fundamental to urban and regional modelling. That is, we have defined modelling problems in relation to urban and regional systems and we have not relied on the use of analogies with other systems. We feel strongly that this is the correct approach, partly because it is most productive, and partly because of the danger of misusing apparent analogies. However, it is now perhaps appropriate in retrospect to examine the concepts of this book in relation to so-called 'social physics' (as the prime source of analogy), on the one hand, and in relation to 'general systems theory', on the other. So, we consider firstly the concepts of social physics; secondly, whether we can rely on analogy to advance in relation to our problems of system dynamics; and, thirdly, the concepts and uses of general systems theory.

7.2 Social analogues of some physical concepts

We begin by exploring simple analogies between concepts in physics and social physics in order to use these as guides to see what certain laws in social physics *might* look like. It is clear from the systems studies already presented in this book that there is going to be a close analogy between cost or utility, and energy. It is important to try to get the signs right. The examples suggest an identification of cost and energy, but since the cost of transport is an estimate of the minimum utility of making trips, we can still have the sign either way. Expenditure on transport in the system is a measure of the amount of money which is being made available in the system for transport, and so is not a bad measure of energy. It is all a matter of getting the definitions correct as in national income accounting, and then applying the correct signs in a consistent way, and these are usually obvious in particular subsystem models. There is actually a similar duality of a sign between energy and work in physics. They are of opposite sign because work is something which is done outside the system on the system. So, we shall try an identification of concepts of the form

$$(\text{costs, utility, money}) = (\text{energy, work}) \ . \tag{7.1}$$

Concepts in physics closely related to the right-hand side of Equation (7.1) are heat and temperature. Heat is a rather special kind of energy which flows between systems in contact. Temperature is a property of a system in equilibrium. Two systems, when placed in contact, form a composite system which is itself in equilibrium if the temperatures of the two separate systems are the same. It is worth discussing briefly a generalisation of the concept of heat due to Jaynes (1957) which makes it less special. We can, given a set of equations each giving the expectation value of a cost or utility, define a type of heat (expenditure or utility flow) and a corresponding 'temperature' in each case.

Consider a generalisation of the system introduced in Section 1.2.2 of Chapter 1. We have probability p_i that a random variable takes value x_i. We know that

$$\sum_i p_i = 1 \tag{7.2}$$

$$E[f_r(x)] = \sum_i p_i f_r(x_i) \tag{7.3}$$

for a number of values of $r = 1, m$ say. [In other words, we know the expectation values for several functions instead of for a single function as in Equation (1.10)]. We can define the entropy S to be

$$S = -\sum_i p_i \ln p_i \tag{7.4}$$

in the usual way. Then the reader can easily check that the p_i which maximises entropy subject to Equations (7.2) and (7.3) can be obtained as follows (and that the method of Section 1.2.2 is a special case of this).

Form the partition function

$$Z(\lambda_1, \lambda_2, \lambda_3, ...) = \sum_i \exp\{-[\lambda_1 f_1(x_1) + ... + \lambda_m f_m(x_m)]\} \tag{7.5}$$

and then[15]

$$p_i = \exp\{-[\lambda_0 + \lambda_1 f_1(x_1) + ... + \lambda_m f_m(x_m)]\}, \tag{7.6}$$

[15] If we define functions f_i, F_i, such that

$$f_i(T_{i'j}) = \begin{cases} 1 & \text{if } i = i' \\ 0 & \text{otherwise} \end{cases}$$

$$F_j(T_{ij'}) = \begin{cases} 1 & \text{if } j = j' \\ 0 & \text{otherwise} \end{cases},$$

and if

$$p_{ij} = \frac{T_{ij}}{T},$$

then, for example, Equations (1.1) and (1.2) of Chapter 1 can be written

$$\sum_i \sum_j p_{ij} f_r(T_{ij}) = \frac{O_r}{T}$$

$$\sum_i \sum_j p_{ij} F_r(T_{ij}) = \frac{D_r}{T},$$

which are in the form of Equation (7.3) above. Equation (1.3) is already in this form. Equation (7.6) can then be used to write down the gravity model immediately! (see Wilson, 1969a, for a detailed statement of this result—the derivation of the gravity model using Jaynes's mathematics directly).

where

$$\lambda_0 = \ln Z \tag{7.7}$$

and the other Lagrangian multipliers $\lambda_1 - \lambda_m$ are determined from the constraint Equations (7.3) which can be written in the form

$$E[f_r(x)] = -\frac{\partial}{\partial \lambda_r} \ln Z \ . \tag{7.8}$$

Note that the maximum entropy value for the distribution is

$$S_{\max} = \lambda_0 + \lambda_1 E[f_1(x)] + \ldots + \lambda_m E[f_m(x)] \ . \tag{7.9}$$

Possible fluctuations can be studied using the variance of the distribution of $f_r(x)$, and this is found to be

$$\Delta^2 f_r(x) = E[f_r(x)^2] - \{E[f_r(x)]\}^2 = \frac{\partial}{\partial \lambda_r^2} \ln Z \ . \tag{7.10}$$

Finally, the function f_r may depend on other parameters $\alpha_1, \alpha_2, \ldots$ as well as x, and it is easily shown that the maximum-entropy estimates of its derivatives are

$$E\left(\frac{\partial f_r}{\partial \alpha_k}\right) = \frac{1}{\lambda_r} \frac{\partial}{\partial \alpha_k} \ln Z \ , \tag{7.11}$$

which will be a useful result for other calculations.

Now, vary the functions $f_r(x)$ in an arbitrary way, so $\delta f_r(x_i)$ may be specified independently for each r and i. Also, let the expectation values of the f_r change in a manner which is also independent, so that there is no relation between $\delta E(f_r)$ and $E(\delta f_r)$. Then the variation in Equation (7.7), using Equation (7.5), can be expressed as

$$\delta\lambda_0 = \delta \ln Z = -\sum_r \delta\lambda_r E(f_r) + \lambda_r E(\delta f_r) \tag{7.12}$$

and so, using Equation (7.9),

$$\delta S = \sum_r \lambda_r [\delta E(f_r) - E(\delta f_r)] \tag{7.13}$$

$$= \sum_r \lambda_r \delta Q_r \ , \tag{7.14}$$

where Equation (7.14) defines Q_r as the rth form of heat. λ_r is seen to be the integrating factor for the rth type of heat, and so is the rth type of temperature. Thus, in the usual case of one constraint equation specifying the expectation value of energy E, the special version of Equation (7.13) has already been given as Equation (1.24).

We can, given any equations giving cost or utility expectation values, define a type of heat and a corresponding temperature. The two kinds of change implicit in Equation (7.13) can easily be illustrated for the transport system. Changes in the network, or the c_{ij}'s, are $E\delta(f_r)$-type changes.

while a change in $\overline{C}$, say due to an overall increase in real income, would be a $\delta E(f_r)$-type change. Since changes are brought about by an interaction between the system of interest which is being modelled, and some other system, this creates the opportunity for heat to flow. We shall return to this example below.

To obtain an example where more than one kind of temperature is involved, consider the elementary utility-location model of Section 6.4.2 of Chapter 6. Suppose that the average cost of housing in zone i is h_i and this is taken as a measure of the benefit of living in that house, and the average wage rate in j (in this simple world) is w_j, also a benefit. c_{ij} is the transport cost as usual.

Then we might take

$$u_{ij} = \alpha h_i + \beta w_j - \mu c_{ij} \tag{7.15}$$

as the utility of living in i and working in j [a form of Equation (6.46)]. [It can easily be checked that this is exactly equivalent to breaking Equation (6.47) into three equations for the different components of utility as described in Chapter 6, and then α, β, and μ are the multipliers associated with these equations.] Then the location model [a variation of Equation (6.48)] becomes

$$T_{ij} = \exp(-\lambda + \alpha h_i + \beta w_j - \mu c_{ij}), \tag{7.16}$$

where h_i, w_j, c_{ij} are different kinds of 'energy' and α, β, and μ are the corresponding 'temperatures'(16). Different kinds of 'heat' will flow in a disequilibrium situation.

We have now had several examples of internal variables for the system, such as expenditure, utility, or population, and we have explored their analogues in physics. We now need external coordinates which will describe urban settlements, and we can then use these to seek to formulate laws and to make predictions about development over time. These external coordinates will not be 'size' in the sense of number of system components, as we have defined them, but are more likely to be physical characteristics of the system. Thus, 'land area' (and hence, implicitly, 'density') is likely to be an external coordinate (of the same nature as volume for a physical system) and 'investment' in certain kinds of capital goods another (so, for example, the c_{ij}'s and hence $\overline{C}$ will be in part a function of investment in transport systems). Work will have to be done to change the values of external coordinates, and it can easily be seen that this is best interpreted as 'investment', which generates utility, or energy, measured by people's expenditure as internal variables.

(16) To obtain Equation (7.16), the Lagrangian multiplier associated with Equation (7.15) has been 'absorbed' into α, β, and μ.

7.3 Disequilibrium and system dynamics

The need to study system dynamics was mentioned earlier. There is also seen a need to study disequilibrium in systems, and here analogues of physical concepts may help. If a physical system is not in equilibrium, then each part of the system may not be at the same temperature. Such systems can be analysed, and there is an obvious analogy with urban systems. It is interesting to consider Harris's (1964) paper in this context, in that his β (our μ and temperature in physics) varies over the population, and in fact has a Γ distribution. This would be a kind of temperature disequilibrium. We now go on to explore system dynamics and attempt to learn about the nature of equilibrium. It was argued in Chapter 1 that one of the advantages of physical analogies in the social sciences is that they may teach us how to study system dynamics. We now examine the available techniques for studying entropy-maximising systems dynamically, and thus for studying disequilibrium.

Firstly, what does it mean to study system dynamics? One customary definition is that a dynamic model contains a variable representing time explicitly, and thus equations of motion can be developed representing the time development of the system. In this sense, thermodynamics, given its usual connotation, should really be termed thermostatics, as the methods used are much more like the comparative static techniques used in economics. The second law of thermodynamics, in physics and social physics, makes statements about directions of change, that is that entropy increases, but does not give an expression for entropy as a function of time. This is what we must now begin to attempt. Physicists are lucky in the sense that a system which is not in equilibrium returns to equilibrium very quickly in most cases, and so a study of the old and new equilibrium position usually suffices. The time taken to return to equilibrium is called the relaxation time, and this will be quite long for some urban systems. This will be explored briefly, using dimensional arguments below. We shall then go on to discuss in turn how disequilibrium can arise, and the type of analysis which is appropriate in each case.

Simple dimensional considerations (ter Haar, 1954, p.390) give the relaxation time of a physical system as

$$\tau \approx d/v \, , \tag{7.17}$$

where d is the mean free path and v the mean velocity. For helium gas under normal conditions this is of the order of 10^{-7} s. If the mechanism of the return to equilibrium in an urban system is the spread of information concerning an innovation, then it is not difficult to estimate that urban relaxation times may vary from hours or days after the provision of a transportation facility to years for major changes in locational opportunities.

What do the classical statistical mechanics' texts tell us about disequilibrium? Tolman (1938, Chapter VI) demonstrates the second law, that entropy increases with time, for a classical system. He also

(Section 107) studies the development of two systems represented by canonical ensembles which are brought into contact, and derives an important result in Equation (107.11). Again, however, he does not study the *process* by which the new equilibrium situation is achieved. ter Haar (1954) goes further. He demonstrates the second law in Chapter II, paragraph 3; studies the coupling of two canonical ensembles in Chapter V, paragraph 6, and also the coupling of two grand canonical ensembles in Chapter VI, paragraph 3. More important still, he studies the process of the return to equilibrium in his Appendix II, for the quantum case. Certain guidelines can be drawn from this Appendix to enable an equivalent classical system to be studied. We believe the really important part of the argument to be the following: that, in studying the process of the return to equilibrium, some *rate* must be supplied which determines how rapidly *transitions* of various kinds can take place. This notion has been developed in a different way, using kinetic theory analogies, in two important papers by Tomlin (1969, 1970).

How can disequilibrium arise? The short answer to this question is: by changing an external coordinate of the system. Possible entity states, such as the expenditure on a trip, *and* the expectation values of such variables can be independently functions of such external variables, and we have the possibility of the two kinds of change outlined in the previous section. There is a very important third way, however, by which disequilibrium can be introduced into an urban system represented by an ensemble which does not appear to be common in physics. That is, a new entity state, or opportunity, may be made available.

To fix ideas, consider urban transport systems and one example of each of the three ways disequilibrium can arise. A gravity model will arise if linkages have costs c_{ij} in the usual way and the mean expenditure on transport is $\overline{C}$. Examples of the three kinds of change are:

(1) a change in one or more of the c_{ij}'s (due to an intracity road improvement or a fare change);
(2) a change in $\overline{C}$ (say, due to a change in income levels);
(3) a bridge may be built linking two previously unconnected areas creating new destination opportunities for travellers (it may be useful to consider that c_{ij} values which were previously infinite have become finite, but this should be a matter of mathematical taste).

Before discussing in more detail the mathematical problems raised in studying the processes of return to equilibrium in each of the above cases, a special kind of 'steady-state' disequilibrium situation should be mentioned. In this case, a disequilibrium situation is achieved *and then maintained* so that the disequilibrium flows of energy flow steadily. Such situations are well known in various branches of physics: the flow of electricity results from a potential difference being maintained, the flow of heat from a temperature difference, and so on. In a single situation, such flows are described by phenomenological equations of the form

$$J_i = L_{ii} X_i \ , \tag{7.18}$$

where J_i is the current, X_i the potential or temperature difference, and L_{ii} a coefficient. In the two physics examples cited, Equation (7.18) represents Ohm's law and Fourier's law, respectively. Many such situations could arise in urban or regional systems. For example, a government may maintain wage differentials between regions to establish certain migration flows. In physics, several such effects can exist simultaneously, and Equation (7.18) is then generalised to

$$J_i = \sum_j L_{ij} X_j \ , \tag{7.19}$$

and it can be shown that

$$L_{ij} = L_{ji} \ . \tag{7.20}$$

Equations (7.19) and (7.20) are known as Onsager's relations and are derived from first principles by ter Haar in his Appendix II. Thus, if a temperature difference is maintained in certain substances, a flow of electricity can occur—the thermoelectric effect. It would be interesting to discover such cross effects for urban systems. There must be several examples, for instance with economic systems. The flow of 'heat' would be the flow of utility or money. A second example to add to the migration one above would be the maintenance of investment incentives to establish certain interregional flows of goods. The two effects are likely to be linked through coefficients like L_{ij} above, the nonzero values of L_{ij} for $i \neq j$, arising because of the obvious interrelations of the two processes. Then the equivalent of something like the thermoelectric effect would be that the maintenance of a wage differential alone would itself generate a flow of goods as well as people.

We now try to outline some of the problems which arise mathematically in representing the process of return to equilibrium. It is simplest to begin with changes of types (1) and (2) defined above. The new equilibrium positions can be found easily, as can the changes in entropy and other quantities. It will be necessary to specify rather carefully the *external* coordinates of the system and the changes in system component values or mean values which can be brought about. It should be possible to define all these quantities in such a way that the thermodynamic laws are satisfied. This may only be so if certain restrictions are placed on the economic and social values implicit in the system. So this in itself could say something interesting about economics.

Let us try to be more explicit. Take the recreational system of Section 4.2 of Chapter 4 so there are n people visiting recreation centres, n_i visiting the ith centre which commits them each to expenditure ϵ_i. $n\epsilon$ is the known total expenditure on recreation. Then, writing in terms of n_i

rather than $p_i = n_i/n$ as in Chapter 4, we have

$$\sum n_i = n \tag{7.21}$$

$$\sum n_i \epsilon_i = n\bar{\epsilon}\,, \tag{7.22}$$

and we can maximise

$$S = -\sum n_i \ln n_i \tag{7.23}$$

subject to Equations (7.21) and (7.22) to obtain

$$\frac{n_i}{n} = \frac{\exp(-\mu\epsilon_i)}{\sum\limits_i \exp(-\mu\epsilon_i)}\,. \tag{7.24}$$

Let α_j be the *external* coordinates of the system. Then we know [cf. Equation (A2.41) in Appendix 2 and adjacent discussion]

$$\mathrm{d}S = \mu n\,\mathrm{d}\bar{\epsilon} + \mu\sum_j \overline{A_j}\,\mathrm{d}\alpha_j\;, \tag{7.25}$$

where

$$\overline{A_j} = E\left(\frac{\partial S}{\partial \alpha_j}\right), \tag{7.26}$$

and are called the external 'forces' corresponding to the external coordinates. They are defined in Equation (A2.38), and following, of Appendix 2. If a change is made in one of the α's (note that $\bar{\epsilon}$ can change independently of the ϵ_i's), then Equation (7.25) gives the change in S [using Equations (7.23), (7.24), and (7.26)]. The new equilibrium positions can be found. However, the second law of thermodynamics states that

$$\mathrm{d}S \geqslant 0\,. \tag{7.27}$$

In physics $\overline{A_j}\,\mathrm{d}\alpha_j$ represents an amount of work done on the system, and Equation (7.27) expresses the fact that the total energy absorbed by the system must be less than, or equal to, the work done. In a socioeconomic system, $\mathrm{d}\alpha_j$ will be a change in a coordinate and A_j something like the investment cost of a unit change. $n\,\mathrm{d}\bar{\epsilon}$ will be a measure of the *utility gain* of the system. Now, it is easy to see that the inequality Equation (7.27) could be infringed if the work done on the system were not properly valued. So this indicates how a reasonable requirement, that a socioeconomic second law should hold, will impose lower bounds on the valuation of certain kinds of investment. If this is followed through, it should enable us to calculate certain kinds of shadow prices.

Finally, a comment on the study of the process of return to equilibrium for changes of this type: ter Haar, in his Appendix II, shows how representations of the external coordinates α_i can be chosen in such a way that Onsager's relations, Equations (7.19) and (7.20), are satisfied, and then J_i, X_i, and the α_j's can be obtained as functions of time. It will be necessary, however, to specify how individuals become aware of the changed opportunities. In physics, the transition rates from one of the old energy levels to new ones can be postulated. In a social system, it is not so easy. Consider a type (1) and type (2) change in turn:

Type (1). In the recreation example, let one ϵ_i change. Then, after time t, it could be assumed that some proportion $\gamma(t)$ of the population were aware of the change and behaved accordingly. That is, $\gamma(t)$ of the population distribute themselves in equilibrium as though the new expenditure levels obtained, and $[1-\gamma(t)]n$ as though they did not. The *total* population is then *not* in equilibrium, though it tends to equilibrium for t greater than some t_0. Entropy would increase continually. The only difficulty here is to be explicit about the mechanism of 'finding out'! How, for example, does one deal with the person who is unaware of the change, who turns up at the centre whose ϵ_i has changed and who then finds that the price has increased? Curiously this situation may be more difficult to analyse than the 'new-level' type (2) situation *where all the old choices* are still available.

Type (2). A change in $\bar{\epsilon}$, say due to increasing income, can be analysed similarly. The most common change of this type will be more analogous to a slow reversible change, $\bar{\epsilon} \rightarrow \bar{\epsilon} + d\epsilon_1, \rightarrow \bar{\epsilon} + d\epsilon_1 + d\epsilon_2$, than to a sudden irreversible change. The problem will be to decide how $d\bar{\epsilon}$ is allocated and how the people who receive it will behave. The mechanism suggested above for type (1) changes can be used, perhaps with $\gamma(t)$ defined in a more complex way as a random variable.

The case which seems most likely to obtain for social systems will be the occurrence of types (1), (2), and (3) changes simultaneously, and so it is important now to study the type (3) situation. We shall also see below that nearly all contact between systems involves the provision of new opportunities for individuals in the population, and so, as the study of laws is often best pursued through the study of contact, such examples should be considered in this example. In fact, some thought shows that methods similar to those outlined above for types (1) and (2) cases can be used. The population has to be divided, at any point in time, into those who are aware of the new opportunities and those who are not.

It is clear that similar methods could be applied to the study of contact between systems. *It is equally clear, and should be emphasised, that other mechanisms than the simple one used here for the spread of information, which is so crucial to the study of disequilibrium, could be postulated and should be investigated.*

A further important part of this research on system dynamics will be the study of isolated systems brought into contact. Such situations are studied in physics to see what form laws take in these circumstances. It is interesting in itself to think out what this could mean for urban systems. One example arises as follows: consider two towns separated by an unbridged river, and thus effectively separated and without contact. Then build a bridge and see what happens. Suppose the transport networks before the bridge is opened are described by $c_{ij}^{(1)}$ and $c_{ij}^{(2)}$. Then the combined prebridge travel time matrix can be written

$$\left(\begin{array}{c|c} c_{ij}^{(1)} & \infty \\ \hline \infty & c_{ij}^{(2)} \end{array}\right).$$

After the bridge has opened, however, the infinite terms become finite with the use of the bridge:

$$\left(\begin{array}{c|c} c_{ij}^{(1)} & c_{ij}^{(A)} \\ \hline c_{ij}^{(B)} & c_{ij}^{(2)} \end{array}\right).$$

Thus, for the combined system, new states $c_{ij}^{(A)}$ and $c_{ij}^{(B)}$ have been introduced. It is not difficult to imagine that, if μ_1 and μ_2 are the temperature parameters of the two isolated systems, then that of the combined system will be μ, where

$$\mu_1 \leqslant \mu \leqslant \mu_2 \,, \tag{7.28}$$

since some trippers in the high μ (congested) system will make substitute trips to the low μ (less congested) system. This will be one consequence of an urban equivalent of the second law of thermodynamics. There will also be a general rise in entropy as required by the second law.

In conclusion, we note that some disequilibrium arises from changes in external variables; system definition is obviously an important topic in the study of dynamics. The system of interest will consist of a number of components and the macrovariables, such as cost and 'temperature' will be, in some form, aggregates of component variables, and so there is a close relation between macro and microdescription of the system of interest. It should also be emphasised that in a particular subsystem, we may only be concerned with a subset of component characteristics. Thus, in Chapter 2, we considered the residents of a city and their journey-to-work behaviour only. (And hence it has been suggested that a necessary precondition of building a dynamic model of entity behaviour may be the building of a dynamic model of the macrovariables such as C.) It is also important to recognise that the external coordinates of the system of interest may be the internal variables of another system (or systems) which interact, and so the problem of specifying variables and external coordinates is closely related to that of subsystem interaction. The main lesson to learn is that, if we force ourselves to be precise and explicit about defining variables and external coordinates of urban and regional systems, then an existing body of mathematics, much of it demonstrated in this book, is available to help us study dynamics.

7.4 Entropy and 'systems' concepts

There has been, and is, considerable interest in what is called 'general systems theory'[17]. The implication of this name is that it is possible to write down a general theory of system behaviour, and then to apply this to, say, urban and regional systems in terms of the context of this book. The basic references are given in a useful paper by Chisholm (1967). A useful general text is that by Klir and Valach (1967). The main concepts of general systems theory are those concepts of social physics which have been outlined and discussed in this book. It has been argued, at least implicitly, that these can be extremely useful. Chisholm, however, concludes his review with the comment: "General systems theory seems to be an irrelevant distraction."

As a general conclusion, this statement is less than fair. (It is probably based on a number of misapplications of the concepts.) There are some obvious advantages to be gained from the general study of the systems. For example, the study of analogies between different systems can often bring about a situation where the study of one system helps the study of another (see Wilson, 1969c). More importantly perhaps, it encourages discipline in the definition of systems, subsystems, their components, and any mutual interactions. If then some general concept is to be applied to a system, a subsystem, or a component, the way in which it is applied will be quite precise. The concept of entropy has a special role in systems theory because it is useful (as can be imagined even from the relatively limited, but still wide, range of application in this book) in the analysis of so many different kinds of systems. This makes it all the more important to be careful about how it is used—otherwise it is easy to seek analogies where none exist. We discuss in turn the need to be precise about the concept of entropy, the way in which probability distributions can be defined, and the notions of closed and open systems. 'Entropy', for example, is not some Platonic property of an urban system which can be discussed without further definition. There is a measure of entropy associated with any probability distribution, and, conversely, whenever the concept of entropy is used it should be used only as the entropy of a probability distribution, which it should be possible to set out explicitly. It is this failure to be explicit which leads to many of the paradoxes with which we are often entertained. Some authors, for example, argue that *self-organising* systems with decreasing entropy are the ones which should interest us. Such paradoxes usually arise because the associated probability distributions have not been carefully defined.

In many cases, probabilities can be clearly defined, at least operationally, in frequency terms, and this has usually been the case with the applications in this paper. However, when the concept of entropy is used more loosely

[17] There are alternative names, such as 'systems analysis', and, for example, though it is not a synonym, 'cybernetics'.

about urban systems, the probabilities which are implicitly involved are often subjective probabilities, for example, of particular decision makers, and are conditional on the state of prior knowledge. Thus the probability distributions in a decision maker's mind, and the system probabilities *about the same set of possible events*, may be different. Great care then has to be exercised to make clear which of these, or which set of these, is associated with the entropy or entropies being discussed. These ideas are discussed by Murphy (1965), and it turns out, for example, that entropies associated with subjective probability distributions of a decision maker about a system can decrease at the same time as the actual system entropy is increasing.

In a similar way, discussions in the literature (for example, Berry, 1967; Woldenberg and Berry, 1967) are often confused because definitions of system boundaries and system constituents are not explicit. In such papers, it is emphasised that the only important and interesting systems are open systems, because only in such systems can entropy be decreased and the resulting 'order' sustained by the supply of energy from the environment. This is really misleading. There may be special cases where this kind of entropy decrease can take place, though it is usually localised and for a short time. An open system is one which interacts with its environment, but then the total system, of open system plus its environment, must be studied. The popular misconceptions about entropy-decreasing systems are well laid to rest in a paper by von Foerster (1960).

In fact, a closed system, explicitly defined, is often a good approximation to equilibrium at a point in time, and this is true for a number of urban systems. It has been shown in this paper, for example, that many gravity models, which are useful for all their deficiencies, can be derived using a microcanonical ensemble—the representation of a closed system. Berry (1967, p.78) says that a closed system will "run down to a state of maximum entropy: any trace of hierarchy would vanish ...". This notion of 'deadness' in a closed system of equilibrium is not a reasonable description of the results which can be derived in using closed systems, as can be seen from many of the examples of this book. Further, the microdistributions remain the same, in equilibrium at a point in time, for open systems, though the constructs of these systems can then be used to study system dynamics. We have shown that it is possible to extend closed-system studies to open-system studies by the introduction of external coordinates. The confusion, and comments such as Berry's above, arise in part because of lack of clarity in the level in which system states are specified. Berry's comment may be true at a macro level (in terms of total 'energy' in some sense), but not true at a more micro level, as with the example of gravity models cited above. However, again the lesson to learn is that it is necessary to be explicit and precise in concept and system definition. It will not usually be useful to talk about an entire urban system and its entropy. An additional specific point can be made in this context. An entire urban regional system is made up of many subsystems. These subsystems may each be in disequilibrium, but tending towards equilibrium. The times taken for each system to gain its equilibrium position are likely to be very different. (This is the problem the econometrician faces by introducing lagged variables.) This fact alone suggests that the application of social physics concepts to an entire urban or regional system, treated as a single system, is likely to give paradoxical results. Thus it could be argued that the work of Curry (1964), Berry (1964), and Ollson (1967), where such broad concepts are used to establish city-size distributions, is not yet convincing.

8

Some conclusions

8.1 General conclusions

The first point to note is that some of the results presented in this book—if not most—could have been derived by more orthodox methods. Some, such as the typical transportation study gravity model, were so derived. The fact remains that most of the results were not derived in this way, and we can safely conclude that entropy-maximising methods provide a useful and practical model-building tool. Typically, these methods are most useful in situations where several rather complex ideas or hypotheses have to be represented in a single theory in an internally consistent way. We also noted that interpretive insights were often gained, and that a path towards the development of dynamic models may have been opened, but these kinds of result are less important, at least at present.

We summarise in the subsections below the conclusions to be drawn about the use of the concept of entropy under a number of headings:

(a) explorations of the concept itself in relation to general systems theory and so-called social physics;
(b) transport;
(c) locational analysis;
(d) other contributions to economics, geography, and planning.

8.2 Explorations of the concept

We have shown in Chapter 1 that there are three (related) ways of viewing the concept of entropy as it is being used in this book. It usually suffices to work with any one of these consistently, and the important point to note is that entropy as an objective system property is yet another use which should be carefully distinguished. Failure to make these distinctions has caused a certain amount of confusion in the literature. We are almost always using the concept in a subjective way as a tool of the analyst rather than as a property of the system.

8.3 Transport

Chapters 2 and 3, and parts of Chapter 5, apply entropy-maximising methods to the development of transport models. Two main conclusions can be drawn. Firstly, the gravity model which has usually been derived using analogies with Newtonian mechanics can be more fruitfully derived using statistical concepts, and, in particular, by maximising entropy. This gives new insights into the traditional model. Secondly, given the new theoretical base, it is possible to elaborate the traditional model in many ways—to recognise that a whole family of spatial interaction models can be developed to represent a variety of transport-flow situations. So, in Chapter 2, we showed how to account for several transport modes, the different behaviour of different person types (particularly car owners and non car owners), and the allocation of trips between alternative routes. In Chapter 3, we showed how the same principles can be used to build a model of the flows of goods, which is internally consistent with a multiregional input–output model. In Chapter 5, two specific and relatively minor questions in relation to transport models were tackled: how to handle external zones and trips across study area boundaries, and how to deal with a situation where parking constraints applied, either because of physical capacity constraints or because of planning policy. Each such development of the traditional transport model is useful in itself, but we can also note that it is easy to integrate the different developments into one model where appropriate.

8.4 Locational analysis

From Chapter 4, we conclude that, in suitable circumstances, spatial interaction models can be used to predict the location of certain activities. One of the advantages of the entropy-maximising framework is that it allows us to be explicit about the nature of the 'suitable circumstances'. Because of the essentially statistical nature of such models, we can expect that they will be most applicable to systems with large numbers of components—for example, to the location of retail activity and to residential location. They probably cannot be applied in the same way in the field of industrial location.

8.5 Other contributions to economics, geography, and planning

The models described in this book obviously contribute to the fields of economics, geography, and urban and regional planning, and likewise the conclusions drawn above. However, there are a number of additional points not mentioned above which can be noted, along with a number of more general comments.

It is always interesting to explore the contribution of a new method of analysis to an old problem, and Chapter 6 is an example of such an exploration in relation to the analysis of utility-maximising systems in economics. It shows that the entropy-maximising method may have a contribution to make as an approximation, either in circumstances where the traditional analysis can be formulated in principle but is mathematically intractable in practice, or, in the more customary sense, when there is imperfect information.

Most of the models presented in this book relate to economics in the sense that the independent variables relate to 'demand' in some sense, and to geography in the sense that they are concerned with the *spatial* distribution of trips and of activities. It has always been very difficult to handle the concept of 'demand at a location' in either economics or geography, and another conclusion which can be drawn is that entropy-maximising methods probably constitute good methods of cutting through some intractable problems raised in the fundamental theoretical analysis of locational demand at least as a first approximation.

The development of models in this field contributes in an obvious way to urban and regional planning. They help to provide the basis for a much-needed predictive capability. There is a sort of converse also: new directions for model-building research can be generated by a study of planning problems which are not at present represented in models. For example, it was a concern with planning problems associated with the housing of low-income families which led the present author to develop the disaggregated residential-location model presented in Chapter 4, and the use of entropy-maximising methods facilitates this kind of disaggregation.

Finally, it should be noted that this book is essentially about model-building *methods*, and less about the models themselves, and it must be acknowledged that, as model builders, we are still only scratching the surface of a set of extremely complex problems.

References

Artle, R., 1961, "On some methods and problems in the study of metropolitan economics", *Pap. Reg. Sci. Ass.*, **8**, 71-87.
Artle, R., 1965, *Studies in the Structure of the Stockholm Economy* (University of Columbia Press, New York).
Batty, M., 1970, "Some problems of calibrating the Lowry model", *Environment and Planning,* **2**, 95-114.
Berry, B. J. L., 1964, "Cities as systems within systems of cities", *Pap. Reg. Sci. Ass.*, **13**, 147-163.
Berry, B. J. L., 1967, *Geography of Market Centers and Retail Distribution* (Prentice-Hall, Englewood Cliffs, N.J.).
Blackburn, A. J., 1969, "On a class of market share functions", Working Paper number 27, Centre for Environmental Studies, London.
le Boulanger, H., Lissarrague, P., 1966, "Le modèle d'equilibre préférential", *Synthèse et Formation,* number 27 (Direction Scientifique de la SEMA, Paris).
Chicago Area Transportation Study, 1960, *Final Report,* volume II.
Chisholm, M., 1967, "General systems theory and geography", *Trans. Inst. Br. Geographers,* **42**, 45-52.
Clough, D. J., 1964, "Application of the principle of maximising entropy in the formulation of hypotheses", *Can. Op. Res. Soc. J.*, **2**, 53-70.
Cordey Hayes, M., 1968, "Retail location model", Working Paper number 16, Centre for Environmental Studies, London.
Curry, L., 1964, "The random spatial economy: an exploration in settlement theory", *Ann. Ass. Am. Geographers,* **54**, 138-146.
Dieter, K. H., 1962, "Distribution of work trips in Toronto", *Proc. Am. Soc. Civ. Engrs.*, **88**, 9-28.
Dorfman, R., Samuelson, P. A., Solow, R., 1958, *Linear Programming and Economic Analysis* (McGraw-Hill, New York).
Edens, H. J., 1970, "Analysis of a modified gravity model", *Transp. Res.*, **4**, 51-62.
von Foerster, H., 1960, "On self-organising systems and their environments", in *Self-organising Systems,* Eds. M. C. Yovitts, S. Cameron (Pergamon Press, Oxford).
Gibbs, J. W., 1902, *Elementary Principles in Statistical Mechanics* (Yale University Press, New Haven).
Goldman, S., 1955, *Information Theory* (Prentice-Hall, Englewood Cliffs, N.J.).
ter Haar, D., 1954, *Elements of Statistical Mechanics* (Constable, London).
Halder, A. K., 1970, "An alternative approach to trip distribution", *Transp. Res.*, **4**, 63-69.
Harris, B., 1964, "A note on the probability of interaction at a distance", *J. Reg. Sci.*, **5**, 31-35.
Henderson, J. M., Quandt, R. E., 1958, *Microeconomic Theory* (McGraw-Hill, New York).
Hotelling, A., 1929, "Stability in competition", *Econ. J.*, **39**, 41-57.
Hyman, G. M., 1969, "The calibration of trip distribution models", *Environment and Planning,* **1**, 105-112.
Hyman, G. M., Wilson, A. G., 1969, "The effects of changes in travel costs on trip distribution and modal split", *J. High Speed Ground Transp.*, **3**, 79-85.
Jaynes, E. T., 1957, "Information theory and statistical mechanics", *Phys. Rev.*, **106**, 620-630.
Kinchin, A. I., 1957, *Mathematical Foundations of Information Theory* (Dover Publications, New York).
Klir, J., Valach, M., 1967, *Cybernetic Modelling* (Iliffe Books, London).
Kuhn, H. W., Tucker, A. W. (Eds.), 1956, *Linear Inequalities and Related Systems* (Princeton University Press, Princeton).

Lakshmanan, T. R., Hansen, W. G., 1965, "A retail market potential model", *J. Am. Inst. Planners,* **31**, 134-143.
Leontief, W., Strout, A., 1963, "Multi-regional input-output analysis", in *Structural Interdependence and Economic Development,* Ed. T. Barna (Macmillan, London).
Lindley, D. V., 1965, *Introduction to Probability and Statistics from a Bayesian Viewpoint* (Cambridge University Press, London).
Loubal, P., 1968, "A mathematical model for traffic forecasting", Ph.D.Thesis, University of California, Berkeley.
Lowry, I. S., 1963, "Location parameters in the Pittsburgh model", *Pap. Reg. Sci. Ass.,* **11**, 145-165.
Lowry, I. S., 1964, "A model of metropolis", RM-4125-RC, Rand Corporation, Santa Monica.
Martin, B. V., Memmott, F. W., Bone, A. J., 1961, "Principles and techniques of predicting future demand for urban area transportation", Research Report number 38, Massachusetts Institute of Technology, Cambridge, Mass.
Mogridge, M. J. H., 1969, "Some factors influencing the income distribution of households within a city region", *Studies in Regional Science,* Ed. A. J. Scott (Pion, London).
Murchland, J. D., 1966, "Some remarks on the gravity model of trip distribution and an equivalent maximising procedure", mimeo., LSE-TNT-38, London School of Economics, London.
Murphy, R. E., 1965, *Adaptive Processes in Economic Systems* (Academic Press, New York).
Neidercorn, J. A., Bechdolt, B. V., 1969, "An economic derivation of the 'gravity law' of spatial interaction", *J. Reg. Sci.,* **9**, 273-282.
Olsson, G., 1967, "Central place systems, spatial interaction and stochastic processes", *Pap. Reg. Sci. Ass.,* **18**, 13-45.
Overgaard, K. R., 1966, *Traffic Estimation in Urban Transportation Planning,* Civil Engineering and Construction Series number 37 (Acta Polytechnica Scandinavia, Copenhagen).
Quandt, R. E., Baumol, W. G., 1966, "The demand for abstract transport modes: theory and measurement", *J. Reg. Sci.,* **6**, 13-26.
Quarmby, D. A., 1967, "Choice of travel mode for the journey to work: some findings", *J. Transp. Econ. Policy,* **3**, 273-314.
Raisbeck, G., 1963, *Information Theory* (MIT Press, Cambridge, Mass.).
Sakarovitch, M., 1968, "The shortest chains in a graph", *Transp. Res.,* **2**, 1-11.
Samuelson, P. A., 1947, *Foundations of Economic Analysis* (Harvard University Press, Cambridge, Mass.).
Sasaki, T., 1968, "Probabilistic models for trip distribution", Paper presented to the Fourth International Symposium on Road Traffic Plan, Karlsruhe.
Schneider, M., 1967, "Direct estimation of traffic volume at a point", *Highw. Res. Rec.,* **165**, 108-116.
Shannon, C., Weaver, W., 1949, *The Mathematical Theory of Communication* (University of Illinois Press).
Spurkland, S., 1966, "Mathematical tools for urban studies", mimeo., Paper presented to Directorate for Scientific Affairs, O.E.C.D., Paris.
Stevens, B. H., 1961, "Linear programming and location rent", *J. Reg. Sci.,* **3**, 15-26.
Stouffler, S. A., 1940, "Intervening opportunities: a theory relating mobility and distance", *Am. Soc. Rev.,* **5**, number 6, 845-867.
Theil, H., 1967, *Economics and Information Theory* (North-Holland, Amsterdam).
Tolman, R. C., 1938, *The Principles of Statistical Mechanics* (Oxford University Press, Oxford.

Tomlin, J. A., 1967, "Mathematical programming models for traffic network problems", Ph.D.Thesis, University of Adelaide.
Tomlin, J. A., Tomlin, S. G., 1968, "Traffic distribution and entropy", *Nature, Lond.*, **220**, 97-99.
Tomlin, S. G., 1969, "A kinetic theory of traffic distribution and similar problems", *Environment and Planning*, **1**, 221-227.
Tomlin, S. G., 1970, "Time dependent traffic distributions", *Transp. Res.*, **4**, 77-86.
Tribus, M., 1969, *National Descriptions, Decisions and Design* (Pergamon Press, Oxford).
Tribus, M., Evans, R., Crellin, C., 1964, "The use of entropy in hypothesis testing", mimeo. paper presented to the Tenth National Symposium on Reliability and Control.
Wilson, A. G., 1966, "Gravity models in the physical and social sciences, I: some comparisons", unpublished note, MAU-N-13, Ministry of Transport, London.
Wilson, A. G., 1967a, "A statistical theory of spatial distribution models", *Transp. Res.*, **1**, 253-269.
Wilson, A. G., 1967b, "Towards comprehensive planning models", unpublished note, MAU-N-72, Ministry of Transport, London.
Wilson, A. G., 1967c, "Use of the Darwin-Fowler method for the derivation of the gravity model", unpublished note, MAU-N-81, Ministry of Transport, London.
Wilson, A. G., 1968, "Modelling and systems analysis in urban planning", *Nature, Lond.*, **220**, 963-966.
Wilson, A. G., 1969a, "Entropy maximising models in the theory of trip distribution, mode split and route split", *J. Transp. Econ. Policy*, **3**, 108-126.
Wilson, A. G., 1969b, "Notes on some concepts in social physics", *Pap. Reg. Sci. Ass.*, **22**, 159-193.
Wilson, A. G., 1969c, "The use of analogies in geography", *Geogr. Analysis*, **1**, 225-233.
Wilson, A. G., 1969d, "Developments of some elementary residential location models", *J. Reg. Sci.*, **9**, 377-385.
Wilson, A. G., 1970, "Advances and problems in distribution modelling", *Transp. Res.*, **4**, 1-18.
Wilson, A. G., Hawkins, A. F., Hill, G. J., Wagon, D. J., 1969, "Calibrating and testing the SELNEC transport model", *Reg. Studies*, **3**, 337-350.
Wilson, A. G., Kirwan, R. M., 1969, "Measures of benefits in the evaluation of urban transport improvements", Working Paper number 43, Centre for Environmental Studies, London.
Woldenberg, M. J., Berry, B. J. L., 1967, "River and central places: analogous systems", *J. Reg. Sci.*, **7**, 129-139.
Wootton, H. J., Pick, G. W., 1967, "A model for trips generated by households", *J. Transp. Econ. Policy*, **1**, 137-153.

Appendix 1. The entropy of a probability distribution

We want a quantity $S(p_1, ..., p_n)$ to represent the 'uncertainty' associated with a probability distribution $p_1, p_2, ..., p_n$. Only three conditions have to be satisfied (Jaynes, 1957):

(1) S is a continuous function of the p_i.

(2) If all p_i are equal,

$$A(n) = S\left(\frac{1}{n}, \frac{1}{n}, \cdots, \frac{1}{n}\right) \tag{A1.1}$$

is an increasing function of n.

(3) Suppose events are grouped in various ways, and let

$$\begin{aligned} w_1 &= p_1 + p_2 + ... + p_k \\ w_2 &= p_{k+1} + ... + p_m, \text{etc.} \end{aligned} \tag{A1.2}$$

Then $p_1 | w_1, p_2 | w_1 ...$ are the conditional probabilities of the events $(x_1, x_2, ..., x_k), ...$. We require that the following composition law be satisfied:

$$\begin{aligned} S(p_1, p_2, ..., p_n) = S(w_1, w_2, ...) &+ w_1 S(p_1 | w_1, p_2 | w_1, ...) \\ &+ w_2 S(p_{k+1} | w_2, ...) + ... \end{aligned} \tag{A1.3}$$

Because of condition (1), we only need determine S for rational values of p_i,

$$p_i = \frac{n_i}{\sum_{i=1}^{n} n_i} \tag{A1.4}$$

where the n_i are integers.

We can view this situation as follows: x_i can occur n_i times out of $\sum_{i=1}^{n} n_i$ equal possibilities. That is, we can consider our 'events' $x_1, x_2, ...$ as themselves composite events out of $n_1, n_2, ...$ equal alternatives. Thus, condition (3) gives

$$S(p_1, p_2, ..., p_n) + \sum p_i S(n_i) = S\left(\sum_{i=1}^{n} n_i\right) . \tag{A1.5}$$

In particular, we can choose all n_i equal to m, so Equation (A1.5) reduces to

$$A(m) + A(n) = A(mn) . \tag{A1.6}$$

It can then be shown that (Kinchin, 1957) the only function which satisfies this and condition (2) is

$$A(n) = k \ln(n) \tag{A1.7}$$

where

$$k \geqslant 0 . \tag{A1.8}$$

Substitute from Equation (A1.7) into Equation (A1.5), to obtain

$$S(p_1, p_2, \ldots, p_n) = k \ln\left(\sum_i n_i\right) - k\sum_i p_i \ln n_i = -k\sum_i p_i \ln p_i \tag{A1.9}$$

using Equation (A1.4).

Appendix 2. Theory of ensembles

A2.1 Introduction

The entropy-maximising method outlined in Chapter 1 and developed in the rest of the book, in effect, assigns equal probability to any state which is not excluded by prior information. The usual results of statistical mechanics, and in particular the Maxwell–Boltzmann distribution, can be obtained by taking $f(x)$ in Equation (1.10) as a possible value of the energy of a system. The method of calculation which is more usually used in statistical mechanics is the ensemble method of Gibbs (see, for example, Gibbs, 1902). An ensemble is an enumeration of possible states of the system which are then again, in effect, assigned equal probability, and expectation values of variables are obtained by averaging over the ensemble. The theory of ensembles is important to social physics to complement the entropy-maximising approach, since it shows how to explore methods of representing different kinds of systems by different kinds of ensembles. The two approaches are, of course, equivalent, but in some circumstances one may be easier to use than the other. In physics, different kinds of ensembles represent different kinds of real-world system and it is worth exploring the physical analogies in some depth to see how to construct theories about urban systems. In physics, if an ensemble represents a particular system of interest, it is called a representative ensemble.

Generally speaking, we shall be concerned with so-called classical (or Boltzmann) statistics, as the components which make up urban systems can be considered to be distinguishable, in contrast to atoms and molecules in physics. This appendix leans on the statistical mechanics texts by Tolman (1938) and ter Haar (1954), though many techniques applied to the quantised case in these books are reworked here for the classical case and applied to a simple social science example. The physicist might consider this to be a regression!

Ultimately, we shall want to take both microscopic views and macroscopic views of urban systems, and this section will show how these are related, to some extent, to different kinds of ensemble. The physicist is usually ultimately interested in the macroscopic view, and the relationship of statistical mechanics to thermodynamics, while the urbanist is at least equally interested, if not more interested, in microdistributions (though macrosocial physics is discussed in Chapter 7).

The general theory of ensembles can be related to the concept of phase space and the equations of motion of the corresponding system. It is worth just exploring this briefly. As stated and implied above, an ensemble is a set of copies of the system of interest, with one copy for each positive state of the system. These states are then assigned equal *a priori* probability. Suppose a system is made up of a number of components (say the particles of a gas in a box, or the individuals making trips in a city). There is then a set of variables whose values completely specify a state of the system. This set of variables is usually made up of subsets, where a single subset consists of variables associated with a component. The minimum number of variables needed to specify a state of the system is the number of degrees of freedom of the system. The space associated with a set of such variables is called phase space (γ space for the whole system involving all the variables or μ space for a set of component variables). Thus a point in phase space represents a possible state of the system and the density of points in phase space can be seen as a probability measure. Variables which describe a system in physics are usually written in canonically conjugate pairs q_i, p_i for $i = 1, N$, where N is usually a large number. Thus $\rho(q, p, t)$ can be defined to be the density of points in phase space for the system of interest, where q stands for all the q_i's and p all the p_i's, and t is time.

There are two reasons to justify a digression to outline how systems are described and analysed in classical physics: firstly, in order to see how one might consider, at some stage, describing a corresponding urban system; secondly, because Jaynes argues that by using the concept of entropy this description, together with equations of motion, *adds nothing to the results of statistical mechanics and thermodynamics*, which can all be obtained using entropy directly. It may well be that something like this is true for urban systems, and we should not be seeking equations of motion directly, but searching for some other method of constructing dynamic models. In classical physics, a state of the system would be specified by a set of coordinates q_i and their corresponding velocities $\dot{q}_i$. The energy of the system (usually considered to be independent of time explicitly for systems of interest, conservative systems) is made up of two components: kinetic energy $T(q, \dot{q})$ and potential energy $V(q)$. Then the Hamiltonian of the system can be formed as the total energy

$$H = T + V. \qquad \text{(A2.1)}$$

However, it turns out to be more convenient to describe the system by canonically conjugate *pairs* of variables, q_i and p_i, where p_i replaces $\dot{q}_i$ and is defined as

$$p_i = \frac{\partial T}{\partial \dot{q}_i}. \qquad \text{(A2.2)}$$

Then, H is written as a function of q and p, and the whole of mechanics can be written in a very condensed way. The equations of motion are

$$\dot{p}_i = -\frac{\partial H}{\partial q_i} \qquad \text{(A2.3)}$$

and

$$\dot{q}_i = \frac{\partial H}{\partial p_i}. \qquad \text{(A2.4)}$$

Now, take any other quantity f which can be expressed as a function of q and p. Its equation of motion is

$$\frac{\mathrm{d}f}{\mathrm{d}t} = \sum_i \frac{\partial f}{\partial q_i}\frac{\partial H}{\partial p_i} - \frac{\partial H}{\partial q_i}\frac{\partial f}{\partial p_i} = \{f, H\} \ , \qquad \text{(A2.5)}$$

where $\{f, H\}$ is called a Poisson bracket and is defined by Equation (A2.5). The expectation value of f is

$$E\{f\} = \int f(q, p)\rho(q, p)\,\mathrm{d}q\,\mathrm{d}p \qquad \text{(A2.6)}$$

at time t, if we assume that phase-space density ρ is defined so that

$$\int \rho\,\mathrm{d}q\,\mathrm{d}p = 1 \ . \qquad \text{(A2.7)}$$

It can also be shown that

$$\frac{\partial \rho}{\partial t} = -\{\rho, H\} \ , \qquad \text{(A2.8)}$$

which has the same form as (A2.5), but with an opposite sign.

An ensemble is, in effect, an enumeration of states of the system, and so, for continuous variation, to define an ensemble is to specify the form of ρ (which also, as pointed out earlier, is a probability density).

Ensembles of interest in physics are so called stationary ensembles, so

$$\frac{\partial \rho}{\partial t} = 0 \ , \qquad \text{(A2.9)}$$

and it then follows that ρ can be a function of constants of the motion only. Thus, if E is the energy of the system of interest,

$$\rho = \rho(E) \ . \qquad \text{(A2.10)}$$

This derivation is often used as the basis of statistical mechanics, and different ensembles are *defined* by specifying the form of $\rho(E)$. It is an open question for urban physics, as to whether appropriate ensembles are stationary.

We can now see what this digression into classical physics has achieved:

(1) It shows how a system in physics is described by canonically conjugate coordinates, how equations of motion are set up in terms of the Hamiltonian (energy expressed as a function of q and p), and how the behaviour of phase-space density can be investigated. We might explore these analogies for urban systems, and gain deeper insights into such systems.

(2) Jaynes has pointed out that most of this description and analysis is not needed to achieve the results of statistical mechanics and thermodynamics, and some might conjecture that this also may be true for urban systems.

We can also note:

(3) The equivalent of the basic mathematics for continuous probability distribution is exactly analogous to the discrete case used by Jaynes; though, after this, we shall only give results for discrete distributions, the extensions to the continuous case are now clear.

Define the entropy of $\rho(q,p)$ as $-\int \rho \ln \rho \, dq \, dp$ and then solve the calculus of variations problem of finding ρ to maximise

$$S = -\int \rho \ln \rho \, dq \, dp \tag{A2.11}$$

subject to

$$\int \rho \, dp \, dq = 1 \tag{A2.12}$$

and

$$\int \rho E \, dp \, dq = E\,. \tag{A2.13}$$

This gives

$$\rho = \exp(-\lambda - \mu E)\,. \tag{A2.14}$$

Note that in classical physics, as outlined here, energy plays a special role through the Hamiltonian. With the information theory approach, which uses entropy rather than energy as the central concept, this is no longer the case. This is particularly useful for thinking about urban systems where we do not expect 'energy' to be as special as it is in physics (cf. Chapter 7).

We can now go on to discuss in turn the different kinds of ensemble, using a simple example, and the discrete formalism.

A2.2 The microcanonical ensemble

It is probably best to define and discuss the different ensembles for a specific example, to fix ideas, and so, firstly, it is necessary to set up the example. We shall use the recreational system set up in Section 4.2 of Chapter 4. Consider a town with n residents who, on a particular weekend, each want to make an excursion to some recreational facility. Suppose the recreational facilities they can use are exclusive to the residents of this town, and that the expenditure incurred by an individual using the ith facility (say travel plus admission costs) is ϵ_i. This expenditure is assumed to be independent of the location of the residence of the individual in the town. All we know is that the average expenditure is $\bar{\epsilon}$, and we know all the ϵ_i's. What is our best estimate of n_i, residents who go to i? Our task in this section is to set up a microcanonical ensemble to describe this system, and we shall see that Jaynes's mathematics as set out in Section 1.2.2 can be used to obtain the solution.

The characteristic of a microcanonical ensemble is that it has a fixed number of entities n and a fixed *total* expenditure $E = n\bar{\epsilon}$. This means that it represents an isolated system, and, as n and $n\bar{\epsilon}$ are the only macroscopic variables, and are constant, we can learn little from this ensemble from the macroviewpoint: all states in the ensemble have the same macrocharacteristics. We can learn more from the microviewpoint by trying to answer the earlier question: what is the likely distribution of the n_i's? Since

$$\sum_i n_i = n \,, \tag{A2.15}$$

then, if we define

$$p_i = \frac{n_i}{n} \,, \tag{A2.16}$$

we have

$$\sum_i p_i = 1 \,, \tag{A2.17}$$

and so p_i behaves like a probability distribution. One of our two pieces of knowledge is represented by Equation (A2.15), and hence Equation (A2.17) in terms of p_i, and our second, on average expenditure, by

$$\sum_i n_i \epsilon_i = n\bar{\epsilon} \,, \tag{A2.18}$$

which can be written in terms of p_i as

$$\sum_i p_i \epsilon_i = \bar{\epsilon} \,, \tag{A2.19}$$

using Equation (A2.16). p_i can be interpreted as the probability of a particular single individual travelling to i. However, we can now define the entropy of this probability distribution by

$$S = -\sum_i p_i \ln p_i \tag{A2.20}$$

and we can now solve our problem using the method of Jaynes. Equations (A2.17), (A2.19), and (A2.20) have the same form as Equations (1.11), (1.10), and (1.12), respectively, and so the most probable distribution is

$$p_i = \frac{n_i}{n} = \exp(-\lambda - \mu\epsilon_i) \,, \tag{A2.21}$$

where

$$\exp\lambda = Z = \sum_i \exp(-\mu\epsilon_i) \,, \tag{A2.22}$$

where Z is called the partition function in the usual way.

It is of interest to study Tolman's (1938) more traditional argument. He defines the vector n_i to be a *condition* of the system, and poses the question: what is the probability density in γ space (system phase space) of a condition? That is, what volume of phase space gives rise to the same condition of the system, and what is the most probable condition? Elements of γ space are made up of products of elements of μ space (entity phase space). It can easily be shown, then, that the volume of γ space corresponding to a condition n_i is proportional to

$$P(n_i) = \frac{n!}{\prod_i n_i!} . \tag{A2.23}$$

If so, the most probable condition is obtained by maximising $\ln P$ subject to the constraint Equations (A2.17) and (A2.19) written in terms of n_i:

$$\sum n_i = n \tag{A2.24}$$

$$\sum n_i \epsilon_i = n\bar{\epsilon} . \tag{A2.25}$$

It can easily be shown, using Stirling's approximation, for large N

$$\ln N! = N \ln N - N , \tag{A2.26}$$

that the result given by Equations (A2.21) and (A2.22) is obtained. Note that $\ln P$ is maximised rather than P because it is more convenient, and because we can obtain the same result by maximising any monotonic function of P. $\ln P$ is closely related to entropy S as defined previously. The advantage of using entropy as the prime concept, as in the first method, is that, although the results obtained are identical, the first method does not rely on Stirling's approximation being applicable.

There is a third way of doing this calculation, which is perhaps worth noting. Take $P(n_i)$ in Equation (A2.23) as being the probability that a condition n_i will occur. We could then evaluate the expectation value of one particular n_i as

$$E(n_i) = \frac{\sum_{n_i} P(n_i) n_i}{\sum_{n_i} P(n_i)} . \tag{A2.27}$$

This expression can be evaluated explicitly using contour integration, and is known as the Darwin–Fowler method. It is demonstrated in Appendix 4 below. Identical results are again obtained. One point of possible interest is that the methods resemble those of stochastic theory, and it may be possible to identify the means of variables which are calculated with the mean values of random variables in some stochastic process.

So, what have we learned from the microcanonical ensemble? Firstly, that macroscopically it does not represent a system of great interest, one which is completely isolated. Secondly, however, there are three appropriate types of calculation, which give useful identical results, when we take a microscopic view.

A2.3 The canonical ensemble

We now construct the canonical ensemble to describe the system defined in the example above. The canonical ensemble is constructed out of the microcanonical ensemble by allowing the total system expenditure the possibility of varying. Thus let E be the total expenditure on recreation, so in Section A2.2 we would have had

$$E = n\bar{\epsilon} \tag{A2.28}$$

as a constant, but now, since we want it to vary, we label total expenditure states by i, so E_i represents a possible value. We assume that we can enumerate these possible values. (It is worth mentioning here that we can consider E_i to be a set of discrete levels without loss of generality. We showed earlier how the mathematics can be carried through in an exactly analogous way for the continuous case.)

In this case, there are alternative macrostates, and, in contrast to its use in Section A2.2, *we now define p_i to be the probability of the whole system having total expenditure E_i*. We shall develop a notation for handling microdistributions based on the quantity $P(n_i)$, the probability of there being n_i individuals in microstate i. Then, for the whole system, we have

$$\sum p_i = 1 \tag{A2.29}$$

$$\sum p_i E_i = \bar{E} \tag{A2.30}$$

$$S = -\sum_i p_i \ln p_i \,, \tag{A2.31}$$

where $\bar{E}$ is the (known) expectation value of E_i. Thus, in the usual way, maximising S subject to Equations (A2.29) and (A2.30), we have

$$p_i = \exp(-\lambda - \mu E_i)\,, \tag{A2.32}$$

where

$$\exp\lambda = Z = \sum_i \exp(-\mu E_i) \tag{A2.33}$$

is the partition function in the usual way. Note that there is a fundamental difference between the partition functions defined in Equations (A2.22) and (A2.23). The former refers to individual expenditure ϵ_i, the latter to system expenditures E_i. Their relationship is discussed further below.

We must now make further explorations to see what kind of system is represented by the canonical ensemble, where total expenditure has possible values E_i, with probability p_i canonically distributed and given by Equation (A2.32). If, in Equation (A2.32), we take logs, then $\ln p_i$ thus obtained can be substituted in Equation (A2.31). This shows that

$$S = \lambda + \mu\bar{E}\,, \tag{A2.34}$$

using Equation (A2.30). Hence the maximum value of the entropy of the system is related to mean total expenditure, the quantity λ which is the logarithm of the partition function, and the parameter μ. Note that μ can in principle be found by solving Equation (A2.30), and so the system is defined at present, and its behaviour determined in this canonical ensemble, by the enumeration of the E_i's, and their mean value $\overline{E}$. (These variables are related to microvariables and, in particular, the enumeration of the ϵ_i's for individuals, but this will be considered later.) The values of these variables will be determined by what can be called the constraints within which the system functions (such as the quality of the transport system connecting the town and the recreational areas). These constraints can be described by coordinates or variables, which we shall call external coordinates (such as the widths of roads, etc.). Let these coordinates be x_i, so that the E_i's and $\overline{E}$ are functions of the x_i's: that is, they will change if external coordinates change, for example if a new road is built. We can now take our study further by studying small changes in the system. Since Equation (A2.29) must be satisfied, in a small change we must have

$$\sum_i \mathrm{d}p_i = 0 , \tag{A2.35}$$

and substitution for p_i from Equation (A2.32) gives

$$\sum_i \exp(-\lambda - \mu E_i)(\mathrm{d}\lambda + \mu \mathrm{d}E_i + E_i \,\mathrm{d}\mu) = 0 , \tag{A2.36}$$

so that, if we take the E_i's to be functions of the external coordinates x_i, we have

$$\sum_i \exp(-\lambda - \mu E_i)(\mathrm{d}\lambda + E_i \,\mathrm{d}\mu + \mu \sum_k \frac{\partial E_i}{\partial x_k} \mathrm{d}x_k) = 0 . \tag{A2.37}$$

Now define

$$X_k^i = -\frac{\partial E_i}{\partial x_k} , \tag{A2.38}$$

where we will call X_k^i the generalised force associated with the external coordinate x_k. [Note that Equation (A2.38) is a special form of Equation (A2.14).] We shall see what it means shortly. Its mean value can be written $\overline{X}_k$, and then Equation (A2.37) can be expressed in terms of mean values as

$$\mathrm{d}\lambda + E \,\mathrm{d}\mu - \mu \sum_k \overline{X}_k \,\mathrm{d}x_k = 0 . \tag{A2.39}$$

However, from Equation (A2.34),

$$\mathrm{d}S = \mathrm{d}\lambda + \mu \mathrm{d}\overline{E} + \overline{E} \mathrm{d}\mu , \tag{A2.40}$$

and we can eliminate λ from Equations (A2.39) and (A2.40) to give $\mathrm{d}S$ as

$$\mathrm{d}S = \mu \mathrm{d}\overline{E} + \mu \sum_k \overline{X}_k \,\mathrm{d}x_k . \tag{A2.41}$$

This now gives us a framework for formulating the laws of behaviour of this macrosystem as represented by a canonical ensemble. We can do this by analogy with the laws of thermodynamics. This will be done very broadly here, and relates to the discussion in Chapter 7. The main purpose of this section is to set up the machinery. The laws of thermodynamics indicate the following properties of the system we have set up, with recreational patterns as an example:

The zeroth law. This states that a system represented by a canonical ensemble is in equilibrium, and that this equilibrium is characterised by the parameter μ. It can easily be shown that if two systems are brought into contact each having this same μ, then the composite system will be in equilibrium with μ unchanged.

The first law. This shows that the energy of a system (in our example, total expenditure) can be increased if work is done on the system (in our example this might be investment, to change some of the external coordinates). It is also indicated that energy can flow from one system to another if they are brought into contact, and such a flow of energy is called the flow of heat. Thus, in our example, if a second 'recreation system' is brought into contact with our present one, some individuals may transfer their expenditure from one to the other, according to the relative values of μ, and this would be equivalent to a flow of heat. The first law then states that if heat is taken into account energy, or expenditure in our case, is conserved.

The second law. This is the most famous of the laws of thermodynamics and is concerned with change. It can be put very simply as

$$\mathrm{d}S \geqslant 0\,, \tag{A2.42}$$

but the best way to discuss it is to give a number of illustrations for our example:
(1) If the system is in isolation, then $\bar{E}$ is constant and the x_k's cannot change (as they can only change by someone *outside* the system investing or doing work). Thus Equation (A2.41) shows that

$$\mathrm{d}S = 0\,, \tag{A2.43}$$

so the equality sign in Equation (A2.42) is satisfied. The entropy tends to some maximum value consistent with $\bar{E}$ and then remains constant.
(2) Consider a change which is made very slowly. In this case, all the work done on the system (investment) is absorbed as energy (people use the new investment and pay for it by increasing their expenditure), so

$$\mu \mathrm{d}\bar{E} = -\mu \sum_k X_k \,\mathrm{d}x_k\,, \tag{A2.44}$$

and

$$\mathrm{d}S = 0\,.$$

Finite changes can be made by a succession of small changes of this kind, and such a change is called a reversible change.
(3) For a more abrupt change, some work will be lost in the form of heat (some investment will not be used and paid for), and in this case the change is irreversible and there is a change in entropy:

$$dS > 0 . \tag{A2.45}$$

(4) We have stated as the basis of the zeroth and first laws that if two systems are brought into contact which are not in equilibrium (different μ's) the composite system will not be in equilibrium. The second law now tells us something about directions of change. Suppose the two systems have equilibrium parameters μ_1 and μ_2. Then the composite system will move to an equilibrium position with an intermediate parameter μ, and there will be an increase in entropy. An example of such a process is presented and discussed in Chapter 7.
(5) It can easily be shown that the system represented by a canonical ensemble is in equilibrium and in contact with the outside world with interactions which maintain the value of μ at a constant level. In our example, since μ is closely related to the overall ease of being admitted to recreational centres, this means that, if our system actually behaves like one represented by a canonical ensemble, then this overall ease is being maintained. This result is less interesting than having the machinery available to analyse the dynamics of system development.
(6) In physics, μ can be shown to be inversely proportional to temperature, and so we can easily see what the social analogue of temperature is.

So we see that by representing a system as a canonical ensemble we can begin to study it dynamically. This has been entirely about the macroview of the system, and we must now study the microimplications of representing a system as a canonical ensemble.

Suppose, as in Section A2.2, that there are n_i people with ϵ_i expenditure on recreation. Remember that this time we defined p_i as a system-state probability, so we want to construct out of it $P(n_i)$, the probability of there being n_i individuals with expenditure ϵ_i. Thus

$$P(n_i) = \sum_k p_k , \tag{A2.46}$$

where the k summation is over all system states k with n_i individuals in individual recreation state i. The calculation is carried out below and we shall see that the most probable microdistribution turns out to be exactly the same as that given by the microcanonical ensemble. We will present the calculation in Section A2.5 below after first discussing the grand canonical ensemble.

A2.4 The grand canonical ensemble

The canonical ensemble, for our example, was constructed as an extension of the microcanonical ensemble by relaxing the condition that the total expenditure was constant. The grand canonical ensemble is a further extension obtained by additionally relaxing the condition that the total number of individuals is constant. So, instead of fixing n, it has a mean value $\bar{n}$. We have to give system-state labels to possible values of n. Let the system label be k, and let N_k be the total number of individuals in state k, and E_k the energy. (Note the use of the capital letter N for this purpose, as we already subscript the small letter with microstate labels. Of course, $\bar{n} = \bar{N}$.) Then, if p_i is again the probability of a *system* state occurring,

$$\sum p_i = 1 \tag{A2.47}$$

$$\sum p_i N_i = \bar{N} = \bar{n} \tag{A2.48}$$

$$\sum p_i E_i = \bar{E} \; . \tag{A2.49}$$

We can use the calculations set out in Section 1.1.2 in Equations (1.10)-(1.13) to show that the most probable distribution is

$$p_i = \exp(-\lambda - \alpha N_i - \mu E_i) \; , \tag{A2.50}$$

where

$$\exp\lambda = Z^* = \sum_i \exp(-\alpha N_i - \mu E_i) \; , \tag{A2.51}$$

and Z^* is now called the *grand* partition function. Equation (A2.31) can be used as in Section A2.3 to give the maximum value of S as

$$S = \lambda + \alpha\bar{N} + \mu\bar{E} \; . \tag{A2.52}$$

Small changes can be investigated in an analogous manner to that of Section A2.3, and the laws of behaviour discussed for a system where the total number of individuals in it can vary. This representation is usually used in physics to study systems with particles of different kinds which can interact. An extension of our example can be used to illustrate this.

Suppose there are three kinds of recreation centre: one offering type of facility A, one offering B, and one offering the combined facilities A-B. There are now three systems A, B, and C (= A-B) which can be aggregated into one total system. The grand canonical ensemble can be used to study the equilibrium properties of the total system. Let N_i^A, N_i^B, N_i^C be the possible total populations of the subsystems, and let E_i^A, E_i^B, E_i^C be the total expenditures. Then it can be shown that, in equilibrium,

$$\frac{\overline{N^A}\,\overline{N^B}}{\overline{N^C}} = \frac{Z_A(\mu) Z_B(\mu)}{Z_C(\mu)} \; , \tag{A2.53}$$

where

$$Z_x(\mu) = \sum_i \exp(-\mu E_i^x) \qquad \text{(A2.54)}$$

is the canonical partition function of subsystem X. This is a social equivalent of the law of mass action, a chemical composition law, and it could be used to study the effects of the provision of recreational centres with mixed facilities. A different application of grand canonical ensembles is given in an earlier paper by the author (Wilson, 1966).

Apart from modelling systems with mixed population types, the structure of the grand canonical ensemble sometimes facilitates computation and allows results to be produced which cannot be calculated in lower-level ensembles. This is useful for microdistributions, where again common results are obtained for our example system, and is illustrated in some depth in Section A2.5 which follows.

A2.5 Microdistributions in canonical and grand canonical ensembles

We now present the formal mathematics needed to obtain microview information for systems represented as canonical or grand canonical ensembles. The question can be posed: why should we bother with this if the final results are identical with those microdistributions which can be obtained more simply with a microcanonical ensemble? The answer is that we shall never be able to use the microcanonical ensemble to represent a dynamic system of interest. Thus, when we finally begin to build dynamic macromodels based on canonical and grand canonical concepts, we shall have to turn to this mathematics to get microdistributions. Since it is difficult to extract and rework some of this from the basic literature, it was thought worthwhile to take this opportunity to present it.

We can now study microdistributions in the simple system we are describing by a canonical ensemble. As defined previously, suppose there are n_i individuals with expenditure ϵ_i in the recreational system. Equation (A2.32) gives p_i, the probability of the system having total expenditure E_i. We want to construct from this the probability $P(n_i)$ that there are n_i individuals with expenditure ϵ_i:

$$P(n_i) = \sum_i p_i \ , \qquad \text{(A2.55)}$$

where the summation is over all system states i, with n_i individuals with expenditure ϵ_i. Note that

$$E = \sum_i n_i \epsilon_i \ , \qquad \text{(A2.56)}$$

and that there are

$$\frac{n!}{\prod_i n_i!} \qquad \text{(A2.57)}$$

regions of phase space (obtained by interchanging distinguishable individuals) with this system expenditure, and so this quantity should be used as a weight to attach to $\exp(-\mu E_i)$ when the summation in Equation (A2.55) is carried out. Thus

$$P(n_i) = \frac{\exp(-\lambda - \mu n_i)n!}{n_i!} \sum_{n_1 + n_2 + n_3 + \ldots = n - n_i} \frac{1}{\prod_{k \neq i} n_k!} \exp[-\mu(n_1 \epsilon_1 + \ldots)] , \qquad (A2.58)$$

and then

$$E(n_i) = \sum_{n_i} P(n_i) n_i = \exp(-\lambda) \sum_{n_1 + n_2 + \ldots = n} \frac{n!}{n_1! n_2! \ldots} \exp[-\mu(n_1 \epsilon_1 + \ldots)] , \qquad (A2.59)$$

where $E(n_i)$ is the expectation value of n_i. The problem is now to evaluate this expression. To do this, note that the partition function Equation (A2.33) can be written in terms of microlevel variables, using the weights Equation (A2.57) as

$$Z = \exp\lambda = \sum_{n_1 + n_2 + \ldots = n} \frac{n!}{\prod_i n_i!} \exp[-\mu(n_1 \epsilon_1 + \ldots)] . \qquad (A2.60)$$

Note that

$$\frac{\partial \ln Z}{\partial \epsilon_i} = \frac{\sum_{n_1 + n_2 + \ldots = n} \frac{n!}{\prod_i n_i!} (-\mu n_i) \exp[-\mu(n_1 \epsilon_1 + \ldots)]}{\sum_{n_1 + n_2 + \ldots = n} \frac{n!}{\prod_i n_i!} \exp[-\mu(n_1 \epsilon_1 + \ldots)]} , \qquad (A2.61)$$

and, so comparing Equations (A2.59) and (A2.61),

$$E(n_i) = -\frac{1}{\mu} \frac{\partial \ln Z}{\partial \epsilon_i} . \qquad (A2.62)$$

However, Equation (A2.60) can be evaluated directly: it is

$$Z = \left[\sum_i \exp(-\mu \epsilon_i) \right]^n , \qquad (A2.63)$$

so, substituting in Equation (A2.62),

$$E(n_i) = \frac{n \exp(-\mu \epsilon_i)}{\sum_i \exp(-\mu \epsilon_i)} , \qquad (A2.64)$$

which is the same result which would have been obtained using a microcanonical ensemble. Note also that this demonstrates the relation between the microcanonical partition function in Equation (A2.22) and the canonical one in Equations (A2.33) and (A2.63). If we write these as Z_{mc} and Z_c respectively, then Equations (A2.22) and (A2.63) show that

$$Z_c = (Z_{mc})^n \ . \tag{A2.65}$$

We can explore the relationship of microcanonical to canonical ensembles one step further: we shall explore the relationship of entropy as defined in a microcanonical ensemble, and in a canonical ensemble. Let these entropies be S_{mc} and S_c respectively:

$$S_c = -\sum_i p_i \ln p_i \tag{A2.66}$$

$$= \lambda + \mu \bar{E} \ , \tag{A2.67}$$

using Equation (A2.32). Note that Equation (A2.67) can also be written

$$S_c = \ln Z_c - \mu \frac{\partial}{\partial \mu} \ln Z_c \tag{A2.68}$$

and similarly

$$S_{mc} = \ln Z_{mc} - \mu \frac{\partial}{\partial \mu} \ln Z_{mc} \ . \tag{A2.69}$$

So, using Equation (A2.65), it can easily be seen that

$$S_c = nS_{mc} \ . \tag{A2.70}$$

Another interesting exercise is to attempt to build up the canonical entropy S_c in Equation (A2.67) from what might be termed expenditure level entropies:

$$S_i = -\sum_{n_i} P(n_i) \ln P(n_i) \ , \tag{A2.71}$$

where $P(n_i)$ is the probability of there being n_i particles with energy ϵ_i. Note that

$$S_i = -E[\ln P(n_i)] \ . \tag{A2.72}$$

However, it can easily be shown that the appropriate calculations are *not* possible because of the restriction that the total number of particles should be n. This idea will be explored further in association with the grand canonical ensemble below, where this restriction on the total number of particles is lifted.

There is a general theorem which further relates the results for microcanonical and canonical ensembles: that any part of an isolated system in equilibrium can be represented by a canonical ensemble (ter Haar, 1954, p.131). In particular, one particle individual in a recreational system can be represented by a canonical distribution with the same parameters as the ensemble which represents the whole system, and this is the result proved more directly in Equation (A2.64) above, as applied to the recreational example.

The next stage in the analysis is to take a microview of the system represented by a grand canonical ensemble, and to show that some of the difficulties we encountered with the canonical ensemble can now be overcome. In Equation (A2.50), i is a state label. We now have to construct out of this p_i particles in the system with energy ϵ_i. By an analogous argument to that which produced Equation (A2.58) in the canonical analysis, we obtain

$$P(n_i) = \frac{\exp[-\lambda-(\alpha+\mu\epsilon_i)n_i]}{n_i!}\sum_{n_1=0 \text{ not } n_i}^{\infty} \cdots \sum_{k\neq i} \frac{n!}{\prod n_k!}\exp\left[-\sum_{k\neq i}(\alpha+\mu\epsilon_k)n_k\right] \tag{A2.73}$$

The essential difference, apart from the form of the probability function p_i, is that the summations on the right-hand side extend over all values of the n_k's and not over those which sum to $n-n_i$ as in Equation (A2.58). Now

$$E(n_i) = \sum_{n_i=0}^{\infty} P(n_i)n_i \ , \tag{A2.74}$$

and again, exactly by analogy with the steps to Equation (A2.62), we have

$$E(n_i) = -\frac{1}{\mu}\frac{\partial\lambda}{\partial\epsilon_i} \ . \tag{A2.75}$$

In this case,

$$\exp\lambda = \sum_{n_i=0}^{\infty} \cdots \sum \frac{n!}{\prod_i n_i!}\exp\left[-\sum_k(\alpha+\mu\epsilon_k)n_k\right] \tag{A2.76}$$

$$= n!\prod_i \sum_{n_i=0}^{\infty} \frac{1}{n_i!}\exp[-(\alpha+\mu\epsilon_i)n_i] \tag{A2.77}$$

by rearranging $\sum$ and $\prod$, and so

$$\exp\lambda = n!\prod_i \exp\{\exp[-(\alpha+\mu\epsilon_i)]\} \tag{A2.78}$$

$$= n!\exp\left\{\sum_i \exp[-(\alpha+\mu\epsilon_i)]\right\} \ . \tag{A2.79}$$

Thus

$$\lambda = \ln n! + \sum_i \exp[-(\alpha+\mu\epsilon_i)] \ , \tag{A2.80}$$

so, substituting for λ in Equation (A2.75),

$$E(n_i) = \exp(-\alpha-\mu\epsilon_i) \tag{A2.81}$$

for the microdistribution, which is the same result again, now with α playing the role previously played by the microcanonical partition function.

The next step is to explore the relation between the system entropy given by Equation (A2.52) and the entropies associated with the probability distribution $P(n_i)$ for each individual expenditure 'level' i. The analogous problem was investigated in Equations (A2.67) and (A2.71) for the canonical ensemble, but the algebra proved too difficult because of the restraint of a fixed number of particles. With this restraint removed, the calculation becomes possible. From Equation (A2.73)

$$\ln P(n_i) = -\lambda - (\alpha + \mu\epsilon_i)n_i + \ln n! - \ln n_i! + \sum_{k \neq i} \exp[-(\alpha + \mu\epsilon_k)] \quad , \qquad \text{(A2.82)}$$

since

$$\sum_{n_1=0}^{\infty} \cdots \sum_{\text{not } n_i} \frac{1}{\prod_i n_i!} \exp\left[-\sum_{k \neq i} (\alpha + \mu\epsilon_k)n_k \right]$$
$$= \exp\left\{ -\sum_{k \neq i} \exp[-(\alpha + \mu n_k)] \right\} \qquad \text{(A2.83)}$$

by analogy with Equation (A2.79). Now substitute for λ from Equation (A2.80):

$$\ln P(n_i) = -\exp[-(\alpha + \mu\epsilon_i)] - \ln n_i! - (\alpha + \mu\epsilon_i)n_i \; . \qquad \text{(A2.84)}$$

Define the entropy of a particular expenditure level to be

$$-\sum_{n_i} P(n_i) \ln P(n_i)$$

and call this S_i. Then S_i can be evaluated by taking expectation values in Equation (A2.84):

$$-S_i = -\alpha\bar{n} - \mu\bar{E}_i - \exp[-(\alpha + \mu\epsilon_i)] - E(\ln n_i!) \; . \qquad \text{(A2.85)}$$

So

$$\sum_i S_i = \alpha\bar{N} + \mu\bar{E} + \lambda - \ln n! + \sum_i E(\ln n_i!) = \lambda + \alpha\bar{N} + \mu\bar{E} - E\left(\ln \frac{n!}{\prod_i n_i!} \right) , \qquad \text{(A2.86)}$$

which is the same as S in Equation (A2.52) except for the last term, which can be interpreted as the interlevel entropy.

Comparing Equations (A2.85) and (A2.86), we can see that we could conveniently define a quantity λ_k such that

$$\lambda_k = \exp[-(\alpha + \mu\epsilon_k)] \qquad \text{(A2.87)}$$

and

$$\exp(\lambda_k) = Z_k \; , \qquad \text{(A2.88)}$$

so Z_k can be called the partition function for the kth 'level'.

The calculations in this section have been carried out for classical distinguishable individuals, as implied by the weight $n!/\prod_i n_i!$ in

Equation (A2.76). We can be more general than this as follows: start from Equations (A2.50) and (A2.51) and write $P(n_i)$ as

$$P(n_i) = \exp(-\lambda - \alpha n_i - \mu\epsilon_i n_i)\sum_l \exp\left(\sum_{k \neq i} -\alpha n_k - \mu\epsilon_k n_k\right) , \tag{A2.89}$$

which reduces to

$$P(n_i) = \exp[-\lambda_i - (\alpha + \mu\epsilon_i)n_i] , \tag{A2.90}$$

where

$$\exp(\lambda_k) = Z_k^* = \sum_{n_k = 0}^{\infty} \exp[-(\alpha + \mu\epsilon_k)n_k] \tag{A2.91}$$

and note that

$$Z^* = \prod_k Z_k^* \tag{A2.92}$$

and Z_k^* is again called the partition function of the kth level. In the case of classical distinguishable entities, we had [though only implicitly in Equation (A2.73)]

$$Z^* = \sum_{n_1}\sum_{n_2} \cdots \frac{n!}{\prod_i n_i!} \exp[-(\alpha + \mu\epsilon_i)n_i] , \tag{A2.93}$$

and we can define

$$Z_k^* = \sum_{n_k} \frac{1}{n_k!} \exp[-(\alpha + \mu\epsilon_k)n_k] . \tag{A2.94}$$

Notice that the $n!$ factor in Equation (A2.93) cannot be 'shared' among levels, and so we cannot make the general relationship Equation (A2.92) hold for classical distinguishable entities. Instead

$$Z^* = n!\prod_k Z_k^* , \tag{A2.95}$$

a relationship which is also implicit in Equations (A2.80) and (A2.87).

Note also that

$$Z^* = \sum_{(N_i)} Z_i \exp(-\alpha N_i) , \tag{A2.96}$$

where Z_i is the canonical partition function, but notice also that α, as pointed out following Equation (A2.81), plays the role of microcanonical partition function, though only in a formal sense.

The general results expressed by Equations (A2.50), (A2.51), and (A2.89)-(A2.92) will be useful if we want to develop models of entities which are not distinguishable. There are then two basic cases: Fermi-Dirac, where n_k can only take the value 0 or 1, and Bose-Einstein, where n_k can take any positive integer value. A possible example for the application of the former is where n_k represents the occupation of a site by a building, say 0 for unoccupied, 1 for occupied.

A2.6 Summary: the types of probability distribution implicit in different kinds of ensemble

An ensemble is a set of copies of a system of interest. Such a system will consist of a set of entities. Ensembles have to be designed so that the sets of copies are built up in such a way that they represent systems of interest, and that means of variables, taken over the ensemble, are the means of variables for the system. Thus ensembles are devices for calculating expectation values. The basic mathematics of canonical distribution has been set out above and this has been applied to three kinds of ensembles: microcanonical, canonical, and grand canonical.

Variables represent either properties of the system (macrovariables) or properties of entities in the system (microvariables). We saw that the three kinds of ensembles have different macroproperties and so represent different kinds of systems, although the canonical is an extension of the microcanonical, and the grand canonical of the canonical. Results which can be proved for a lower-level ensemble can also be proved in a higher-level one, though not necessarily *vice versa*. We also saw that results for microvariables are common to the three kinds of ensemble, though certain calculations are facilitated at higher levels.

It is worth just reviewing briefly the kinds of probability distributions, at macrolevels and microlevels, which can be constructed in the different kinds of ensemble:

(1) Microcanonical: at the macrolevel, there is no probability distribution because all the macrovariables are fixed. At the microlevel, we find the most probable distribution of occupancy numbers n_i, and this turns out to be a negative exponential distribution. The quantity

$$p_i = \frac{n_i}{n} \tag{A2.97}$$

behaves like a probability distribution and can be interpreted as the probability that a *single* entity shall be in state i.

(2) Canonical: at the macrolevel, we now have the probability p_i that the system will have total expenditure E_i, that is, will be in system state i. At the microlevel, we do not calculate n_i directly now, but the probability distribution $P(n_i)$, the probability that there should be n_i individuals with individual expenditure ϵ_i. There are calculation difficulties, especially in relation to the entropy of the distribution $P(n_i)$ which arise because the maximum value of n_i is n.

(3) Grand canonical: the situation is similar to the canonical one, except that n_i, in the distribution of $P(n_i)$, can take any value, and this facilitates computations.

Appendix 3. The intervening-opportunities model and some variants

A3.1 Introduction

A second approach to trip distribution uses the intervening-opportunities model. Interzonal impedance does not appear explicitly in this model, but possible destination zones away from an origin zone i have to be ranked in order of increasing impedance from i. A notation is needed to describe this. Let $j_\mu(i)$ be the μth destination zone in this rank order away from i; $j_\mu(i)$ will be referred to simply as j_μ in cases where it is clear to which i it refers. The intervening-opportunities model was first developed by Stouffler (1940) in a simple form, assuming that the number of trips from an origin zone to a destination zone is proportional to the number of opportunities at the destination zone, and inversely proportional to the number of intervening opportunities. The underlying assumption of the model is that the tripper considers each opportunity, as reached, in turn, and has a definite probability that his needs will be satisfied. The model will be derived here in the form developed by Schneider, originally for use in the Chicago Area Transportation Study (1960). To see how the basic assumption operates, consider a situation in which destination zones are rank-ordered away from an origin zone as defined earlier. Let U_{ij_μ} be the probability that one tripper will continue beyond the μth zone away from i. Suppose there is a chance L that an opportunity will satisfy this single tripper when it is offered. Then, to the first order in L,

$$U_{ij1} = 1 - LD_{j1}$$

where D_{j1} is the number of opportunities in the zone j_1, nearest to i. Then, combining successive probabilities multiplicatively,

$$U_{ij2} = U_{ij1}(1 - LD_{j2})$$

$$U_{ij3} = U_{ij2}(1 - LD_{j3}) ,$$

and so on. In general,

$$U_{ij_\mu} = U_{ij_\mu -1}(1 - LD_{j_\mu}) . \tag{A3.1}$$

This equation can be written

$$\frac{U_{ij_\mu} - U_{ij_\mu -1}}{U_{ij_\mu -1}} = -LD_{j_\mu} . \tag{A3.2}$$

Let A_{j_μ} be the number of opportunities passed up to and including zone j_μ. Then

$$D_{j_\mu} = A_{j_\mu} - A_{j_\mu -1} \tag{A3.3}$$

and Equation (A3.2) can be written

$$\frac{U_{ij_\mu} - U_{ij_\mu -1}}{U_{ij_\mu -1}} = -L(A_{j_\mu} - A_{j_\mu -1}) . \tag{A3.4}$$

This equation can be written, with the assumption of continuous variation [18], as

$$\frac{\mathrm{d}U}{U} = -L\,\mathrm{d}A \ , \tag{A3.5}$$

which integrates to

$$\ln U = -LA + \text{constant} \ ,$$

so that

$$U_{ij_\mu} = k_i \exp(-LA_{j_\mu}) \ , \tag{A3.6}$$

where k_i is a constant. But

$$T_{ij_\mu} = O_i(U_{ij_\mu -1} - U_{ij_\mu}) \ , \tag{A3.7}$$

where T_{ij_μ} is the number of trips from i to the μth destination away from i, for a total of O_i trips originating at i. Substitution from Equation (A3.6) to Equation (A3.7) gives

$$T_{ij_\mu} = k_i O_i \left[\exp(-LA_{j_\mu -1}) - \exp(-LA_{j_\mu})\right] \ , \tag{A3.8}$$

and this is the usual statement of the intervening-opportunities model.

Note that k_i can be chosen so that the resulting matrix T_{ij} satisfies the constraint Equation (1.1):

$$\sum_j T_{ij} = k_i O_i \left[1 - \exp(-A_{jN})\right] = O_i$$

where N is the total number of zones. Since $\exp(-LA_{jN})$ should be very small, k_i will be very nearly 1 for each i. The constraint Equation (1.42) on the total number of trip attractions cannot be satisfied, however, within the model structure itself, but if actual D_j's are known, the matrix can be adjusted by the same balancing process as that implied by Equations (1.7) and (1.8).

[Thus, in general, if there is a matrix T_{ij}^* and it is required to transform it to a matrix T_{ij} whose columns and rows sum to O_i and D_j, then this can be done by the transformation

$$T_{ij} = A_i B_j T_{ij}^* \ , \tag{A3.9}$$

where

$$A_i = O_i\left(\sum_j B_j T_{ij}^*\right)^{-1} \tag{A3.10}$$

$$B_j = D_j\left(\sum_i A_i T_{ij}^*\right)^{-1} \ , \tag{A3.11}$$

[18] It has been pointed out to me by Andrew Evans that there are mathematical difficulties associated with the transition from discrete variation to continuous and back again; however, here I am only presenting the usual derivations from the literature, and the difficulties will not be discussed further.

and these equations can be seen to reduce to Equations (1.7) and (1.8) for the gravity model. This process is accomplished in practice by factoring rows and columns successively by D_j/D_j^* and O_i/O_i^*, where the D_j^* are the column sums, and O_i^* the row sums, of the matrix reached after the immediately preceding operation. This is the balancing process which can be applied, if required, to the intervening-opportunities model-trip matrix.]

In the next section, we show how the intervening-opportunities model can be derived from entropy-maximising principles.

A3.2 Maximum-entropy derivation

The intervening-opportunities model was derived in the traditional way in the introduction above and its main equation was derived as Equation (A3.8). It is of some interest to attempt to derive this using the entropy methodology, since, if this is possible, the gravity and opportunities models are related by this common base and can be compared in a new light.

Using variables defined in Section A3.1 above, it is also possible to define in addition

$$S_{ij_\mu} = O_i U_{ij_\mu} \tag{A3.12}$$

as the number of trips from i continuing beyond the μth-ranked zone away from i. Note that, since

$$T_{ij_\mu} = S_{ij_{\mu-1}} - S_{ij_\mu} , \tag{A3.13}$$

the variables S_{ij_μ} define the new system as a possible alternative to T_{ij_μ}. To derive the opportunities model, the new method is applied to the variables S_{ij_μ} [19]. Thus, if S is the total number of states for a given distribution $\{S_{ij_\mu}\}$, then the maximand will be

$$\frac{S}{\prod\limits_{ij} S_{ij_\mu}!} \, .$$

It is now necessary to establish appropriate constraints. As seen earlier, the opportunities model does not have a constraint on trip attractions of the form of Equations (1.2), but does need a constraint on trip generations of the form of Equation (1.1). For the variables S_{ij_μ}, the constraint strictly analogous to Equation (1.1) is the inequality

$$S_{ij} \leqslant O_i , \tag{A3.14}$$

as there cannot be more trips continuing beyond a point from i, than

[19] In what follows, I have implicitly assumed that the S_{ij_μ}'s are independent variables: Richard Allsop has pointed out to me that they are unlikely to be independent if the T_{ij}'s are independent. However, I have thought it still useful to present this derivation, since even with a possible mathematical weakness, it may prove stimulating.

originally set out from i. If these are summed over j_μ to obtain a constraint of the form of Equation (1.1), the resulting equation is

$$\sum_{j_\mu} S_{ij} = k_i' O_i \ , \tag{A3.15}$$

where k_i' is some constant, and $1 \leqslant k_i' \leqslant N$, where N is the total number of zones. Finally, a constraint analogous to the gravity-model cost constraint, Equation (1.3), is needed.

The main assumption of the intervening-opportunities model, as commonly stated, is that the number of trips between i and j is determined by the number of opportunities at j, and varies inversely as the number of intervening opportunities. This gives the clue for the cost constraint: to use intervening opportunities as a proxy for cost. Thus, if S_{ij_μ} trips are to be made beyond j_μ, then these will incur costs greater than those for trips which have been made to nearer zones. Suppose, then, for trips from i, we take the number of opportunities passed as a measure of the cost of getting so far. Thus the minimum cost for the remaining trips beyond j_μ is $A_{j_\mu} S_{ij_\mu}$. If this is summed over j_μ and then over all origin zones i, this gives a function which behaves in some ways like a total-cost function, and the corresponding constraint is

$$\sum_i \sum_{j_\mu} A_{j_\mu} S_{ij_\mu} = C \ . \tag{A3.16}$$

Since, as can be derived from the definitional Equation (A3.3),

$$A_{j_\mu} = \sum_{n=1}^{\mu} D_{jn} \ , \tag{A3.17}$$

and, as can be derived from the definition of S_{ij_μ},

$$S_{ij_\mu} = \sum_{n=\mu+1}^{N} T_{ijn} \ , \tag{A3.18}$$

where N is the total number of zones, it can easily be seen that the coefficient of T_{ij_μ} in the summation in Equation (A3.16) is [substituting for S_{ij_μ} from Equation (A3.18)]

$$(\mu - 1)D_{j1} + (\mu - 2)D_{j2} + \ldots + D_{j_{\mu-1}} \ , \tag{A3.19}$$

and so the opportunities passed contribute to the cost associated with a particular element of the trip matrix weighted by the number of times they have been 'passed' or 'have intervened'.

Now, maximising

$$\frac{S}{\prod_{ij_\mu} S_{ij_\mu}!}$$

subject to the constraint Equations (A3.15) and (A3.16), introducing

Lagrangian multipliers $\lambda_i^{(1)}$ and L for these constraints, we find the most probable distribution occurs when

$$-\ln S_{ij_\mu} - LA_{j_\mu} - \lambda_i^{(1)} = 0\,,$$

so

$$S_{ij} = \exp(-LA_{j_\mu} - \lambda_i^{(1)})\,. \tag{A3.20}$$

$\lambda_i^{(1)}$ can be obtained in the usual way by substituting from Equation (A3.20) into Equation (A3.15):

$$\exp(-\lambda_i^{(1)}) = \frac{k_i' O_i}{\sum\limits_{j_\mu} \exp(-LA_{j_\mu})}\,, \tag{A3.21}$$

and so, writing

$$k_i = \frac{k_i'}{\sum\limits_{j_\mu} \exp(-LA_{j_\mu})} \tag{A3.22}$$

$$S_{ij_\mu} = k_i O_i \exp(-LA_{j_\mu}) \tag{A3.23}$$

and, using Equation (A3.13), we have

$$T_{ij_\mu} = k_i O_i\, [\exp(-LA_{j_\mu\text{-}1}) - \exp(-LA_{j_\mu})]\,, \tag{A3.24}$$

which is identical with Equation (A3.8). Thus the main equation of the intervening-opportunities model has been obtained using the new method.

This derivation has been made at the expense of using a rather strange cost-constraint Equation (A3.16), and assuming a cost of getting from i to j_μ implied by Equation (A3.19). Perhaps this is an argument in itself for preferring the gravity model to the intervening-opportunities model.

A3.3 A distribution model of gravity type which uses intervening opportunities as a measure of cost

As a final example of the application of the new method, consider the distribution model which is obtained if intervening opportunities are used as a measure of cost, but not weighted in the form of Equation (A3.19). This is perhaps a more plausible assumption.

Note that j_μ, as originally defined, is properly a function of i and should be written as $j_\mu(i)$. This assumption now proposed is

$$c_{ij} = A_{j_\mu(i)}\,. \tag{A3.25}$$

This cost can now be substituted in either of the two gravity models derived above, the so-called conventional model described by Equations (1.6)–(1.8), or the so-called single-competition-term model described by Equations (3.12)–(3.13) with $f(c_{ij})$ as $\exp(-\beta c_{ij})$. Thus, substituting for c_{ij}

from Equation (A3.25), the double-competition-term 'gravity/opportunity' model is

$$T_{ij_\mu} = a_i b_{j_\mu} O_i D_{j_\mu} \exp(-\beta A_{j_\mu (i)}) \tag{A3.26}$$

$$a_i = \left[\sum_{j_\mu} b_{j_\mu} D_{j_\mu} \exp(-\beta A_{j_\mu (i)}) \right]^{-1} \tag{A3.27}$$

$$b_j = \left[\sum_i a_i O_i \exp(-\beta A_{j_\mu (i)}) \right]^{-1} . \tag{A3.28}$$

Small a's and b's are used for the balancing factors to avoid confusion with the A_{j_μ}'s.

The single-competition-term 'gravity/opportunity' model is

$$T_{ij_\mu} = a_i O_i D_{j_\mu} \exp(-\beta A_{j_\mu (i)}) \tag{A3.29}$$

$$a_i = \left[\sum_{j_\mu} D_{j_\mu} \exp(-\beta A_{j_\mu (i)}) \right]^{-1} . \tag{A3.30}$$

If it could be argued that Equation (A3.25) represents a better account of cost than Equation (A3.19), then the models represented by Equations (A3.26)–(A3.28) and Equations (A3.29)–(A3.30) may give better answers than the conventional intervening-opportunities model. A test of these new models would be welcomed.

Appendix 4. Use of the Darwin-Fowler method for the derivation of the gravity model

A gravity model was derived by a maximum-probability method in Section 1.2 of this book. The method maximised $\ln T!/\prod_{ij} T_{ij}!$ subject to a number of constraints on the variables $\{T_{ij}\}$. It was stated in an earlier paper (Wilson, 1967a), as a conjecture, that it would be possible to produce an analogue of the Darwin–Fowler method in statistical mechanics (cf. ter Haar, 1954, for a description of the method) which would calculate *average* values of T_{ij}, $\overline{T}_{ij}$, rather than the most probable. This method does not depend on any approximation and, reassuringly, leads to the same answer. The purpose of this appendix is to present this derivation for the standard transport gravity model. *The maximum probability method of Section 1.2.1 remains the most convenient method in practice for setting up models of particular systems.*

The main assumption of Section 1.2.1 is that the probability of the state T_{ij} occurring is proportional to $W(\{T_{ij}\})$ which, dropping the factor $T!$ can be written

$$W(\{T_{ij}\}) = \prod_{ij} \gamma(T_{ij}) \ , \tag{A4.1}$$

where

$$\gamma(T_{ij}) = \frac{1}{T_{ij}!} \tag{A4.2}$$

(thus defining the function γ for later use). In the usual case, the matrix T_{ij} must satisfy the constraint Equations (1.1)–(1.3), which are repeated here for convenience:

$$\sum_j T_{ij} = O_i \tag{A4.3}$$

$$\sum_i T_{ij} = D_j \tag{A4.4}$$

$$\sum_i \sum_j T_{ij} c_{ij} = C \ . \tag{A4.5}$$

Now, instead of finding the T_{ij} which maximises $W(\{T_{ij}\})$, we find the average value $\overline{T}_{ij}$, defined by

$$\overline{T}_{ij} = \frac{\sum_c T_{ij} W(T)}{\sum_c W(T)} \tag{A4.6}$$

where $W(T)$ [a shorthand for $W(\{T_{ij}\})$] is the probability of state $\{T_{ij}\}$ and T_{ij} is the value of the (i, j) matrix element in this state. $\sum_c$ denotes summation over all states $\{T_{ij}\}$ which satisfy the constraint Equations (A4.3)–(A4.5) and $W(T)$ is defined by Equations (A4.1) and (A4.2).

More generally,

$$\overline{T_{ij}^p T_{mn}^q} = \frac{\sum\limits_{\mathrm{c}} T_{ij}^p T_{mn}^q W(T)}{\sum\limits_{\mathrm{c}} W(T)} \ . \tag{A4.7}$$

The Darwin–Fowler method is a means of evaluating these summations explicitly.

Define a *generating function*:

$$F(x_i, y_i, z;\ c_{ij}) = \sum_{\text{all } \{T_{ij}\}} W(T) \prod_i x_i^{\sum\limits_j T_{ij}} \prod_j y_i^{\sum\limits_i T_{ij}} z^{\sum\limits_i \sum\limits_j T_{ij} c_{ij}} \ . \tag{A4.8}$$

Then, to evaluate $\sum\limits_{\mathrm{c}} W(T)$, where the summation is over all $\{T_{ij}\}$ satisfying the constraint Equations (A4.3)–(A4.5), we have to pick out the terms in the sum defined by Equation (A4.8) whose x_i exponent is O_i, whose y_j exponent is D_j, and whose z component is C. This can be done by complex integration: using Cauchy's theorem applied to each variable in turn, we obtain

$$\sum_{\mathrm{c}} W(T) = \left(\frac{1}{2\pi i}\right)^{2N+1} \oint \mathrm{d}x_1 \oint \mathrm{d}x_2 \ldots \oint \mathrm{d}y_1 \oint \mathrm{d}y_2 \ldots \oint \mathrm{d}z$$

$$\times \prod_i x_i^{-O_i - 1} \prod_j y_j^{-D_j - 1} x^{-C-1} F(x_i, y_j, z;\ c_{ij}) \ , \tag{A4.9}$$

where N is the number of zones.

More generally, we can similarly evaluate the summation Equation (A4.7) by

$$\sum_{\mathrm{c}} T_{ij}^p T_{mn}^q W(T) = \left(\frac{1}{2\pi i}\right)^{2N+1} \oint \mathrm{d}x \ldots \oint \mathrm{d}y \ldots \oint \mathrm{d}z \times \prod_i x_i^{-O_i - 1} \prod_j y_j^{-D_j - 1} z^{-C-1}$$

$$\times \left(\frac{1}{\ln z} \frac{\partial}{\partial c_{ij}}\right)^p \left(\frac{1}{\ln z} \frac{\partial}{\partial c_{mn}}\right)^q F(x_i, y_j, z;\ c_{ij}) \ . \tag{A4.10}$$

All integrations are counterclockwise along a closed contour around the origin in the complex planes of x_i, y_j, and z. We have defined our integrand with singularities in x_i, y_j, and z at values of the exponents we are interested in: O_i, D_j, and C.

Now note that F can be written in the form

$$F(x_i, y_j, z;\ c_{ij}) = \prod_{ij} f(x_i y_j z^{c_{ij}}) \ , \tag{A4.11}$$

where

$$f(\eta) = \sum_{n=0}^{\infty} \gamma(n) \eta^n = \mathrm{e}^{\eta} \tag{A4.12}$$

using the function γ defined in Equation (A4.2). It will be useful later to have the generating function in this form.

To evaluate the integrals, let

$$F(x_i, y_j, z;\ c_{ij})x_i^{-O_i-1} = \exp[O_i g_i(x_i)] \ . \qquad \text{(A4.13)}$$

This simply defines a function g_i, and we can let

$$I_i(x_i) = \frac{1}{2\pi i}\oint dx_i \exp[O_i g_i(x_i)] \ . \qquad \text{(A4.14)}$$

Then, it is known that

$$I_i(x_i) = \exp[O_i g_i(x_i^0)]\left[\frac{1}{2\pi O_i g_i''(x_i^0)}\right]^{1/2} , \qquad \text{(A4.15)}$$

where x_i^0 satisfies

$$\frac{\partial g}{\partial x_i} = 0 \ , \qquad \text{(A4.16)}$$

which is, in terms of F,

$$x_i \frac{\partial F}{\partial x_i} - (O_i + 1)F = 0 \ . \qquad \text{(A4.17)}$$

It can be shown that the second factor of Equation (A4.15) is much smaller than the first [essentially N compared with $\exp N$, as could be checked explicitly using Equations (A4.11), (A4.12), and (A4.13)] and, since we shall, in the end, only be using $\ln F$, and hence $\ln I_i$, this factor can be neglected. Also, 1 can be neglected compared with O_i in Equation (A4.17). So

$$I_i(x_i) = \exp[O_i g_i(x_i^0)] = F(x_i^0, y_j, z;\ c_{ij})(x_i^0)^{-O_i-1} \qquad \text{(A4.18)}$$

where x_i^0 is the solution of

$$x_i \frac{\partial F}{\partial x_i} = O_i F \ . \qquad \text{(A4.19)}$$

Each integral can be evaluated similarly and combined in Equation (A4.9) to give

$$\sum_{c} W(T) = \prod_i (x_i^0)^{-O_i-1} \prod_j (y_j^0)^{-D_j-1} z^{-C-1} F(x_i^0, y_j^0, z^0;\ c_{ij}) \ , \qquad \text{(A4.20)}$$

where x_i^0 satisfies Equation (A4.19), and, similarly, y_j^0 and z^0 satisfy

$$y_i \frac{\partial F}{\partial y_j} = D_j F \qquad \text{(A4.21)}$$

$$z \frac{\partial F}{\partial z} = CF \ . \qquad \text{(A4.22)}$$

It can similarly be shown that the integral Equation (A4.10) is

$$\sum_{c} T_{ij}^{p} T_{mn}^{q} W(T) = \prod_{i} x_i^{-O_i-1} \prod_{j} y_j^{-D_j-1} z^{-C-1} \left(\frac{1}{\ln z^0}\frac{\partial}{\partial c_{ij}}\right)^p \left(\frac{1}{\ln z^0}\frac{\partial}{\partial c_{ij}}\right)^q$$

$$\times F(x_i^0, y_j^0, z^0;\ c_{ij}) \ . \qquad \text{(A4.23)}$$

So, substituting from Equations (A4.20) and (A4.23) into Equation (A4.7), we have

$$\overline{T_{ij}^{p} T_{mn}^{q}} = \frac{1}{F}(\ln z)^{-p-q} \left(\frac{\partial}{\partial c_{ij}}\right)^p \left(\frac{\partial}{\partial c_{mn}}\right)^q F(x_i, y_j, z;\ c_{ij}) \ . \qquad \text{(A4.24)}$$

It is easy to see that, for this F in the form given in Equations (A4.11) and (A4.12), this expression vanishes unless either $p = 0$ or $q = 0$. Thus the most general result is

$$\overline{T_{ij}^{p}} = \frac{1}{F}(\ln z)^{-p} \frac{\partial^p}{\partial c_{ij}^p} F \ , \qquad \text{(A4.25)}$$

and, in particular, when $p = 1$,

$$\overline{T_{ij}} = \frac{1}{F \ln z} \frac{\partial F}{\partial c_{ij}} \ . \qquad \text{(A4.26)}$$

Equations (A4.11) and (A4.12) show that

$$F = \prod_{ij} \exp(x_i y_j z^{c_{ij}}). \qquad \text{(A4.27)}$$

Thus, noting that if $F(c) = \exp[f(c)]$, and $f(c) = Az^c$,

$$\frac{\partial F}{\partial C} = \exp(Az^c) Az^c \ln z \qquad \text{(A4.28)}$$

we can write Equation (A4.26) as

$$\overline{T_{ij}} = x_i^0 y_j^0 (z^0)^{c_{ij}} \qquad \text{(A4.29)}$$

where x_i^0, y_j^0, and z^0, it will be recalled, satisfy Equations (A4.19), (A4.21), and (A4.22) respectively. We now explore these equations in more detail.

We have F given explicitly in terms of the x's, y's, and z in Equation (A4.27). Note that

$$\ln F = \sum_{i} \sum_{j} x_i y_j z^{c_{ij}} \qquad \text{(A4.30)}$$

(Note here and onwards that we only use $\ln F$, not F.) Then

$$\frac{\partial \ln F}{\partial x_i} = \frac{1}{F}\frac{\partial F}{\partial x_i} = \sum_{j} y_j z^{c_{ij}} \ ; \qquad \text{(A4.31)}$$

so, substituting in Equation (A4.19),

$$x_i = \frac{O_i}{\sum_j y_j^0 (z^0)^{c_{ij}}} . \tag{A4.32}$$

Similarly,

$$y_j = \frac{D_j}{\sum_i x_i^0 (z^0)^{c_{ij}}} , \tag{A4.33}$$

and

$$\frac{C}{z} = \sum_{ij} c_{ij} x_i^0 y_j^0 (z^0)^{c_{ij}-1} . \tag{A4.34}$$

Equation (A4.34) is analogous to the cost-constraint equation in the previous work, and, as usual, cannot be solved for z^0. We can define a β so that

$$\exp(-\beta c_{ij}) = (z^0)^{c_{ij}} \tag{A4.35}$$

(i.e. $\beta = -\ln z^0$). Finally, write

$$x_i^0 = A_i O_i \tag{A4.36}$$

$$y_j^0 = B_j D_j \tag{A4.37}$$

(thus defining A_i and B_j to replace x_i^0 and y_j^0), and then Equations (A4.29), (A4.32), (A4.33), and (A4.35) can be combined with Equations (A4.36) and (A4.37) to give

$$\overline{T}_{ij} = A_i B_j O_i D_j \exp(-\beta c_{ij}) \tag{A4.38}$$

$$A_i = \left[\sum_j B_j D_j \exp(-\beta c_{ij})\right]^{-1} \tag{A4.39}$$

$$B_j = \left[\sum_i A_i O_i \exp(-\beta c_{ij})\right]^{-1} , \tag{A4.40}$$

which are the usual equations, obtained in Section 1.1 as Equations (1.6)-(1.8). So the average value $\overline{T}_{ij}$ is the same as the most probable value. [It is now apparent that the differential Equations (A4.19), (A4.21), and (A4.22), solved at Equations (A4.32)-(A4.34), are simply restatements of the usual constraint equations.] This completes the derivation of the gravity model by use of the Darwin-Fowler method.

Author index

Subject index

For Product Safety Concerns and Information please contact our EU
representative GPSR@taylorandfrancis.com
Taylor & Francis Verlag GmbH, Kaufingerstraße 24, 80331 München, Germany

www.ingramcontent.com/pod-product-compliance
Ingram Content Group UK Ltd.
Pitfield, Milton Keynes, MK11 3LW, UK
UKHW011016180926
475701UK00004B/776

* 9 7 8 0 4 1 5 6 9 6 3 1 9 *